Introductory College Chemistry

Laboratory Experiments

Irina Rutenburg
Paris Svoronos
Pedro Irigoyen

Queensborough Community College

KENDALL/HUNT PUBLISHING COMPANY
4050 Westmark Drive   Dubuque, Iowa 52002

Copyright © 2006 by Kendall/Hunt Publishing Company

ISBN13: 978-0-7575-2968-9

Kendall/Hunt Publishing Company has the exclusive rights to reproduce this work, to prepare derivative works from this work, to publicly distribute this work, to publicly perform this work and to publicly display this work.

All rights reserved. No part of this publication may be reproduced, stored in a retrieval system, or transmitted, in any form or by any means, electronic, mechanical, photocopying, recording, or otherwise, without the prior written permission of Kendall/Hunt Publishing Company.

Printed in the United States of America
10 9 8 7 6 5 4

Contents

Introduction to Chemistry Laboratory: Check-in, Safety Film and Discussion — v

Experiment 1	Physical Properties of a Substance: Density	1
Experiment 2	Physical Properties of a Substance: Melting and Boiling Points	9
Experiment 3	Chromatography	19
Experiment 4	Formula of a Hydrated Salt	27
Experiment 5	Chemical Properties of a Substance: Chemical Reactions	33
Experiment 6	Electrical Conductivity of Aqueous Solutions: Electrolytes and Nonelectrolytes	41
Experiment 7	Molar Mass of a Volatile Gas	49
Experiment 8	Determination of a Solution's Concentration by Visible Spectrophotometry	57
Experiment 9	pH Determination of Solutions	65
Experiment 10	Determination of a Solution's Concentration by Titration	75
Experiment 11	Titration of Buffers	83
Experiment 12	Nuclear Chemistry: Radioactivity	91
Experiment 13	Qualitative Analysis	99
Experiment 14	Chemical Equilibrium and Le Chatelier's Principle	105

Introduction to Chemistry Laboratory

I. Check-in, Safety Film and Discussion

The American Chemical Society Safety film must be seen by every student at the beginning of the semester. No student will be allowed to perform any experiment unless he/she has passed the safety quiz after viewing the film.

Eye Protection in Chemical Laboratory

Eye protection is not just a departmental requirement in chemical laboratory; it is a **STATE REGULATION**. State law requires that safety glasses (goggles) must be worn in chemical laboratory **at all times**.

Wearing goggles at all times means:

- Students have to wear goggles no matter whether they are working, talking to the instructor, copying something from the blackboard, or writing something in the lab notebook.

- Students are not allowed to lift the goggles up when recording date from instruments or talking to somebody. Goggles must be worn on the bridge of the nose, not on the forehead, neck, or shoulder.

- If students need to leave the room, they should walk to the exit with the goggles still in front of their eyes and have them placed on again when they return to the lab. If students need to adjust their goggles, they should leave the lab with the goggles on, adjust them, and put them on again before they re-enter the room.

  ❏ Students should wash their hands before touching their goggles, especially when adjusting or taking them off.
  ❏ Cleaning up after the experiment is over should be done while still wearing the goggles!
  ❏ Should the goggles be kept in the lab drawer, they should be well packed and not left loose and in contact with the equipment and possible chemical contamination.
  ❏ Sunglasses and regular vision glasses cannot be used as goggles. Goggles must be worn on top of vision goggles if the student needs to wear vision goggles in the lab.
  ❏ Wearing contact lenses is not allowed in the lab (even if the student wears goggles) as the vapors of volatile chemicals may be "sandwiched" between the contact lens and the eye.

Procedure for Accidental Eye Injury

If a chemical accidentally gets into the eye(s), the student must first inform the instructor immediately. The eye(s) should be washed immediately with a lot of water several times. Medical attention should also be immediately sought.

Safety Procedures in the Chemical Laboratory

- Doors to a chemical laboratory room must be left open at all times when laboratory classes are in session.
- Only students registered for the lab in session are allowed to enter the laboratory. No other students or visitors are allowed to enter the laboratory at any time without the instructor's permission.
- Students are not allowed to enter the lab before their instructor's arrival.
- No students are allowed to enter the stockroom located in S-435.
- Students should read the experiment prior to coming to class.
- Students must be in class on time and should carefully listen to the instructor's introduction of the experiment.
- Students may not be allowed to perform an experiment if they arrive late and miss the introduction.
- Students are not allowed to start the experiment before the instructor gives permission to do so.
- Students must stop working after the instructor has requested work to stop.
- Students are not allowed to perform any unauthorized experiments.
- Students should always be aware that the instructor may make important additional announcements or experimental modifications at any time.
- Students must know the experiment's waste chemicals' disposal directions.
- Students are responsible for cleaning their benches and glassware, and washing and returning all equipment and chemicals to the carts.
- Hands must be washed with soap before leaving the lab.
- Paper, paper towels, filter paper, litmus paper, paper clips, rubber bands, and all other solid objects must not be placed into sink or troughs. Instead they should be placed in the available trash bins.

Safety Behavior in the Chemical Laboratory

- No shorts, sandals, high thin heels, bulky clothes, shawls, scarves, hanging jewelry, or loose hair styles are allowed. Only sneakers or flat shoes are allowed in the laboratory.
- Long hair should be tied in the back with a rubber band.

- Backpacks and coats should not be kept on the benches where the experiments are performed. Instead they should be kept in designated areas.
- No food or drinks are allowed in the lab under any circumstances.
- Tasting, smelling, or touching of chemicals is strictly not allowed.
- No chemicals, glassware, or any other equipment may be taken away from the lab without the knowledge of the instructor.
- Test tubes should not be directed at anybody when they are heated or when chemicals are being mixed in them.
- Hot glassware or metalware must be cooled first before further handling.
- Extra chemicals left over or produced from the experiment should not be placed back into bottles, but disposed as directed by the instructor.
- Liquids should be dispensed or transferred using funnels.

Accidents and Emergencies

- Any fire in the room must be announced immediately to the instructor, and the whole class must be evacuated immediately in an orderly manner.
- The eyes of a student should immediately be washed thoroughly with a lot of water if they accidentally come in contact with chemicals.
- If a chemical gets into the mouth, it should be flushed with a lot of water.
- If a chemical gets in touch with skin, it must be washed with a lot of water.
- If a thermometer is broken, the instructor must be notified immediately.
- Broken glass should not be handled by hand. Instead it should be handled with a paper towel and disposed of in the appropriately designated container.
- Both instructors and students should be aware of the location and purpose of eye wash, safety showers, fire extinguishers, fire blankets, and buckets.
- The instructor must be informed immediately of any accident described previously and, in addition, medical attention should be sought as the result of any accidents.

Students' Grades and Obligations

The students' grade will not be based solely on their lab reports, but also on:

- Their attendance, which is mandatory. **Departmental policy dictates that three inexcused absences will lead to an automatic WU. Only the instructor can decide if the absence is excused. In order to pass the course you must pass both the lecture and laboratory parts of the course.**
- Their coming to class on time to listen to the introduction of the experiment;
- Their work and performance in class;
- Their abiding by the rules of safety and cleaning.

The instructor has the right and obligation if a student violates any of the rules:
- To lower the student's grade;
- To have the student removed from any particular class;
- To have the student removed from the course.

II. Physical and Chemical Properties of a Substance

One of the major goals of chemistry is to determine how various substances differ from each other. Each substance has characteristic **physical properties** that identify it in a way similar to a description of a person's eye and hair color, complexion, height and weight, fingerprints, and so on. Similarly, a substance has its own characteristics; such as color, taste, and smell. A substance is also characterized by the way its matter is naturally packed, which is defined as its **density.** A substance can exist in the solid, liquid, or gaseous state, all of which are known as its **physical states.** A substance can undergo a **physical change,** which is **reversible.** Thus, a solid may be converted to a liquid and the liquid may be solidified; the liquid may be boiled to become a gas and the gas may be condensed into a liquid. Every substance undergoes these changes at a certain temperature and pressure. Some substances conduct electricity while others do not: **electron conductivity** is what describes this physical property. Some substances may interact with magnetic fields and are called **paramagnetic,** while others do not and are called **diamagnetic.**

Substances can also be characterized by the ways in which they interact with each other. Some substances mix together and are said to be **miscible;** others do not mix and they are referred to as **immiscible.** Another way of discussing miscibility of two substances is to say that one substance **dissolves** in another. **Solubility** is one of the physical properties of a substance, which indicates the extent to which one substance dissolves in another. This is an important property because most substances are found as mixtures with other substances. The various processes that lead to the separation of substances from mixtures is also a very important area of chemistry.

Each substance can also be characterized by its **chemical properties,** which are defined by the chemical reactions it undergoes. A chemical reaction is also called a **chemical change.** During a chemical change, one substance changes into another. The study of chemical properties of substances lies at the very heart of chemistry.

Experiment 1

Physical Properties of a Substance: Density

Introduction

Matter occupies space and has weight mass. Mass is a measure of the amount of matter in any object, from which the object is made. Volume is the space that the object occupies. **Density** of the substance is defined as the mass of a quantity of the substance that occupies a unit volume, or as the ratio of mass to volume:

$$d = m/V,$$

where

$$d \text{ is density}$$
$$m \text{ is mass and}$$
$$V \text{ is volume}$$

Density is usually expressed as grams (g) per milliliter (mL) for solids and liquids. Note that 1 mL = 1 cm^3. Density is expressed in grams (g) per liter (L) for gases. Thus;

$$d \text{ (solids or liquids)} = g/mL \text{ or } g/cm^3$$

and

$$d \text{ (gases)} = g/L.$$

Density provides a way to assign a numerical value to this physical property of matter. The higher the number, the denser the substance. Table 1.1 lists a number of substances with the corresponding density at 25°C.

Density is characteristic of the substance under specific temperature and pressure (for gases) conditions. Substances that are denser than water sink. The opposite holds for substances that are less dense than water. Looking at Table 1.1, one concludes that gasoline floats because it is less dense than water. Mercury, on the other hand, is denser than water and will sink. A gas like helium will enable a balloon to float up to the ceiling or out the window simply because it is less dense than air.

TABLE 1.1 Densities of Common Substances

Substance	Density (g/mL)
Cork	0.22–0.26
Gasoline	0.75–0.95
Ethyl Alcohol	0.79
Human fat	0.94
Ice (at 0°C)	0.917
Water (at 4°C)	1.000
Table sugar	1.59
Aluminum	2.70
Iron	7.86
Nickel	8.90
Lead	11.34
Mercury	13.6
Gold	19.31

Also:

Air (at 25°C)	1.185 g/L
Helium (at 25°C)	0.164 g/L

The first thing that will be done in this laboratory session is the familiarizing of the student with a computer program called **"Bridging to the lab."** This program will illustrate how to collect data and calculate the density of substances. This user-friendly program is simple to follow and will lead the student along with common operations such as entering data, drop, drag, and click.

The second part will involve the actual, physical determination of the density of various substances in the solid, liquid, and gas phase (state).

Procedure

I. Computer Simulation Experiment

1. Start your computer and sign in.
2. Place the cursor on the **"Bridging to the Lab"** icon and double click.
3. From the menu, pick **"Measuring Density: Is the artifact authentic."**
4. The first window will have the objectives and a list of all the sections.
5. Move the cursor over **"What You Should Know"** (on the lower left-hand corner) to see what you must be familiar with in order to perform this simulation.
6. When you are ready go on, click the **"Start"** button (lower right-hand corner) to begin.
7. Follow the instructions and perform all tasks before moving on by clicking the "Next" button (lower right-hand corner). **Note: the program will not let you move on until you correctly perform all tasks that are called for in that section.**
8. Complete all sections **Introduction, Experimental Setup, Collecting Data, Summary,** and **Self-Test.**

9. The last window will identify all the sections you have completed. Show the screen to your instructor before moving on.
10. Close the program, shut down the computer and begin to perform the actual experiment as follows.

II. Actual Determination of Density

> ### Safety Tips
> Review from the first week: Goggles and handling liquid chemicals.
> Follow your professor instructions on how to dispose of the liquid chemicals.

I. Density of an Unknown Solid (metal bar)

1. Obtain a metal bar and record its color on the data sheet (item A, section I).
2. Weigh the metal bar and record its mass on the data sheet (item B, section I).
3. Place about 50 mL of water into a 100 mL graduated cylinder, read the exact volume, and record it on the data sheet (item C, section I).
4. Slowly slide (**DO NOT DROP**) the metal bar in the graduated cylinder, which is inclined at a 45°C.
5. Record the new volume in the graduated cylinder and place it on the data sheet (item D, section I).
6. Calculate the density of the metal bar and record it in the data sheet (item F, section I).
7. Identify the unknown metal by comparing your calculated density (item F, section I) to the values that appear in Table 1.2 (item G, section I).

TABLE 1.2 Densities of Various Metals

(Your unknown is one of the listed possibilities)

Substance	d (g/mL or g/cm^3)
Aluminum	2.70
Tin	7.28
Iron	7.86
Copper	8.92
Lead	11.3

II. Density of an Unknown Liquid

1. Obtain an unknown liquid and record its number on the data sheet (item A, section II).
2. Weigh an empty 10 mL graduated cylinder and record its mass on the data sheet (item B, section II).
3. Use a funnel to **carefully** place 5–10 mL of the unknown liquid into the graduated cylinder and weigh the graduated cylinder with the liquid. Record on the data sheet (item C, section II).
4. Carefully read the exact volume of the liquid in the graduated cylinder and record it on the data sheet (item E, section II).
5. Calculate the density of the liquid and record it on the data sheet (item F, section II).

III. Density of Air

Though air is a mixture of gases, the same procedure would be used for any other gas whether pure or impure.

1. Obtain the gas densitometer and the vacuum pump from your instructor.
2. Open the valve on the densitometer and attach the hose from the vacuum pump to the valve opening.
3. Start evacuating the air by pumping approximately 20 times. Close the valve and disconnect the hose.
4. Weigh the evacuated densitometer and record the mass on the data sheet (item A, section III).
5. Open the valve of the densitometer and allow the air to fill the vacuum (you will hear a hissing sound of air entering into the evacuated flask).
6. Weigh the densitometer again and record the mass on the data sheet (item B, section III).
7. Record the volume of the densitometer by reading its size as inscribed on the glass (item D, section III).
8. Calculate the density of air and record it on the data sheet (item E, section III).

Date: _____

Class: _____

Name: _____

Experiment 1—Data Sheet

Physical Properties of a Substance: Density

I. Density of Unknown Solid (metal bar)

Measurements and Observations

A. Color of the metal bar _____

B. Mass of the metal bar _____ g

C. Volume of water in the graduated cylinder (before the metal bar is inserted) _____ mL

D. Volume of water in the graduated cylinder (after inserting the metal bar) _____ mL

Calculations

E. Volume of water displaced by the metal bar in the graduated cylinder = volume of metal bar (D−C) _____ mL

F. Density of metal (B/E) _____ g/cm^3

Conclusion

G. After comparing the calculated density to the densities that appear in Table 1.2, my unknown metal is probably _____

II. Density of Unknown Liquid

Measurements and Observations

A. Unknown number _____

B. Mass of empty graduated cylinder _____ g

C. Mass of graduated cylinder and liquid _____ g

D. Volume of liquid in the graduated cylinder _____ mL

Calculations

E. Mass of liquid in the graduated cylinder (C−B) _____ g

F. Density of the unknown liquid (E/D) _____ g/mL

III. Density of Air

Measurements and Observations

A. Mass of the evacuated densitometer _____ g

B. Mass of the densitometer with air _____ g

Calculations

C. Mass of air in the densitometer (B–A) _____ g

D. Volume of the densitometer _____ 0.50 L

E. Density of air (C/D) _____ g/L

Questions and Problems

(Show all calculations)

1. Which is denser, cotton or gold? Which is heavier: 1 cm³ of cotton or 1 cm³ of gold?

2. Coal and diamond are two different forms of carbon. Which is denser?

3. Why does ice float on water?

4. A student measured a mass of a metal bar as 57.21 g. The student filled a graduated cylinder with water and determined the volume of water to be 48.60 mL. Then the student placed the metal bar into the graduated cylinder and determined the new volume to be 53.70 mL. Calculate the density of the metal and identify its name from the choices that appear in Table 1.2.

5. A chemist has a 9.50 mL sample of rubbing alcohol (isopropyl alcohol). What is the mass of the sample? What additional information do you need in order to calculate the mass?

6. What is the mass of 25 mL of water (measured at 4°C)? (*HINT:* use the data that appear in Table 1.1).

Experiment 2

Physical Properties of a Substance: Melting and Boiling Points

Introduction

Solid, liquid, and gas comprise the three physical states (phases) of matter. Every substance can exist in a different state under different conditions of temperature and pressure. For example, water may be encountered as ice, liquid water, or gas (steam or vapor). When the conditions of temperature and pressure change, a substance may undergo a physical change as described here:

Melting Point is defined as the temperature at which both the solid and liquid phases are at equilibrium. Thus, water at 0°C may exist as a solid (ice) and a liquid (water).

Boiling Point is defined as the temperature at which both the liquid and gas phases are at equilibrium. Thus, water at 100°C may exist as a liquid and gas (steam or vapor).

Temperature is usually measured in degrees **Celsius** (°C). Other temperatures include **Kelvin** (K) and **Fahrenheit** (°F). The conversions between the different scales are as follows:

$$F = 1.8 \,(°C) + 32°C = (°F - 32)/1.8 \quad \text{and} \quad K = 273.15 + °C$$

Note: To convert a temperature from °F to K, one needs to go through the following sequence: °F → °C → K.

Melting and/or boiling points (together with other physical properties) are used to identify an unknown substance. An impure substance's melting point will always be lower than that of the pure substance and will have a wide temperature range. On the other hand, impurities increase the boiling point, but less significantly.

In this experiment the boiling point of an unknown liquid and the melting point of an unknown solid.

Safety Tips

Review from the first week: Goggles and handling chemicals. Follow your professor's instructions on how to dispose the liquid and solid chemicals.

Procedure

I. Actual Determination of Boiling and Melting Points

a) Boiling Point of a Liquid

1. Obtain an unknown liquid and record its number on the data sheet.
2. Assemble the boiling point setup as seen in Figure 2.1.
3. Place 10–20 drops of the unknown liquid and a one-end sealed capillary tube into the 6 mm test tube.
4. Lower this setup into the center of the 250 mL beaker, making sure that the liquid in the test tube is well below the water level. The test tube must be at least 1 inch above the bottom of the beaker. At this point, check with your instructor that the set up is appropriately assembled.
5. Slowly start heating the water bath until a steady stream of bubbles emerges out of the open end of the capillary tube. At this point, turn off the heat.
6. As the water bath temperature decreases the rate of bubble formation decreases. Record the temperature at which the last bubble is released.
7. Compare the boiling point of your unknown to the boiling points reported in Table 2.1 and determine the name of the unknown liquid. Record the name of your unknown on the data sheet.

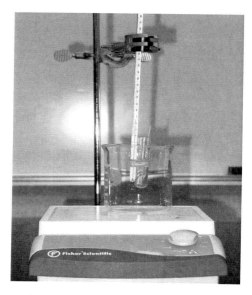

Figure 2.1 Boiling point setup

b) Melting Point of a Solid

1. Obtain an unknown solid substance and record its number on the data sheet.
2. Place a small amount of the unknown substance on a watch glass.
3. Obtain a capillary tube that has one sealed end. Press the open end of the capillary tube into the substance so that a small amount of crystals is forced into the tube.
4. Invert the capillary tube and tap the sealed end of the capillary tube on a hard surface in order to collect the unknown in the bottom sealed end of the tube.
5. Insert the capillary tube in the melting point machine and identify the melting point of the unknown solid as demonstrated by your instructor.
6. Determine the melting point and compare it to the possibilities in Table 2.2.
7. Identify your unknown and place it on the data sheet.

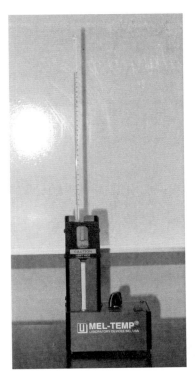

Figure 2.2 Melting point setup

12 Experiment 2 Physical Properties of a Substance: Melting and Boiling Points

II. Computer Interface Determination of Boiling and Melting Points

a) Calibrating the Temperature Sensor

1. Connect the interface to the computer and the correct sensor(s) to the interface. Turn on the interface and then the computer.
2. Click the "Start" button of Windows XP. Go to "Programs," move over to the interface program, and click on it. This will start the interface program.
3. Click on "Calibrate" of the interface program. The sensor calibration Menu will open. Pick the sensor that needs to be calibrated.
4. Obtain a thermometer that will measure the (−10°C) −110°C temperature range.
5. Connect the temperature sensor to the interface unit itself.
6. Obtain a 100 mL beaker, fill it with distilled water to about ¾ its volume, and heat to about 70°C.
7. Click on the "Temperature" sensor of the sensor calibration menu.
8. The temperature sensor window will open, asking for two temperatures to be entered. The first temperature will be "room temperature," which you can read off your thermometer. Type the value in the provided space.
9. Place the sensor in the beaker (from step 6) that contains the hot water. Use your thermometer to read the exact temperature of the hot water and enter it in the provided space.
10. Close the calibration window. The sensor is now calibrated.
11. After calibrating, set up the program to plot temperature (on y-axis) and time in seconds (on the x-axis).
12. Collect data every 10 seconds and plot.

b) Measuring the Boiling Point of a Liquid

1. Place 4 mL of your unknown liquid into the test tube into the microboiling point set up described previously (Figure 2.3). Add one (1) boiling chip. Do not forget to record your unknown number.

Figure 2.3 Computer interface boiling point setup

2. Suspend the temperature sensor in the liquid. Do not let the sensor touch the walls or bottom of the test tube.
3. Start heating the beaker containing the water.
4. Turn the computer on and collect data until the highest temperature is reached, which will remain constant. This is the boiling point of your solution.
5. Print the graph and record your name and unknown liquid number.
6. Identify your unknown by looking at Table 2.1.

Equipment

Boiling Point:
- Temperature Probe
- Computer Interface
- Test tube
- Cork
- Hot Plate
- Ring Stand
- 400 mL Beaker
- Clamp

Melting Point:
- Spatula
- Weigh Paper
- Capillary Tube

Questions and Problems

(Show all calculations)

1. What is the effect of impurities on the boiling point of a compound?

2. What is the effect of impurities on the melting point of a compound?

3. A student measured the melting range point of an unknown solid to be 70.9–71.3 °C.
   a) Is the substance pure?

   b) Can you identify the substance from Table 2.2?

   c) If yes, what is the substance?

4. A student measured the melting point range of an unknown solid to be 72.4–78.6°C.
   a) Is the substance pure?

   b) Can you identify the substance?

   c) Which substances that appear in Table 2.2 can be excluded?

5. A student measured the boiling point of an unknown liquid to be 82.9°C. After consulting Table 2.1, this liquid is _____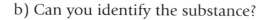

6. Convert 20.0°C (room temperature) to °F.

7. Convert
   a) 20.0°C to K
   b) −10.0 °C to °F
   c) −10.0°C to K
   d) 70.0°F to °C
   e) 70.0°F to K
   f) 300 K to °C

Experiment 3

Chromatography

Introduction

Most substances exist as mixtures of several compounds (components). **Chromatography,** a most useful method, was originally used to separate colored mixtures of substances. Nowadays, even colorless components may be separated by this technique. The general principle of chromatography involves the separation of the mixture components via the use of a **stationary** phase, which is usually a solid in powder form. The components interact differently with the stationary phase and travel on it at different rates. Generally, the greater the interaction between the stationary phase and the component, the slower the traveling. The movement along the stationary phase is achieved via the **mobile** phase (often a solvent), which separates the components along the stationary phase.

This experiment involves two types of chromatography. **Thin layer chromatography** uses a plate as the stationary phase, with the components traveling up the plate from bottom to top by means of a solvent. In **column chromatography,** a packed column is used as the stationary phase, with the substances traveling from top to bottom and separately collected at the bottom as solutions in a solvent (the mobile phase).

Thin Layer Chromatography (TLC)

A plate on which the stationary phase (in the form of powder) is evenly pressed is spotted at one end with a concentrated solution of the mixture of the components to be separated. This end of the plate is allowed to come in contact with the solvent that serves as the mobile phase. The solvent travels along the plate by capillary action and dissolves the spot on its way. The individual components travel together but at different rates by means of the solvent that distributes them on the plate. The separated components appear as spots of different colors (if the original mixture is colored) at different distances from the end of the plate on which the mixture was originally spotted. Thin layer chromatography is generally used for small samples of mixtures. The method of resolution of the components is generally used for identification purposes, such as in forensic science.

The distances traveled by each component as well as the solvent are measured at the end of the experiment. The ratio (R_f) of the distance traveled by a component ($d_\#$) to the distance traveled by a solvent (d_s) is characteristic (and constant) of each component with respect to a particular solvent:

$$R_f = d_\# / d_s$$

Column Chromatography

Column chromatography is used when one or more substances in a mixture need to be isolated in relatively large quantities, such as in pharmaceutical chemistry. A column is suspended and filled with the stationary phase (a powder, such as alumina or silica gel) and solvent (the mobile phase) is added to keep the stationary phase wet. The mixture is then applied at the top and the components are separated (eluted) through the bottom of the column according to their interaction with the stationary phase. It is very important that the column not be allowed to dry at any time; otherwise, channels will be formed and the separation will be negatively affected. Therefore, a continuous addition of the solvent is necessary to keep the stationary phase wet.

Safety Tips

Review from the first class:
1. Wearing goggles
2. Handling liquid and powdered chemicals
3. Handling capillary glass tubes, disposable glass pipets, etc.

Follow your professor's instructions on:
1. The use and disposal of dyes, solvents, and powdered chemicals
2. Disposal of capillary tubes and disposable pipets

Procedure

Thin Layer Chromatography (TLC)

Special items: Development chamber and lid, capillary tubes, tweezers, scissors, TLC sheets (stationary phase), solvents, and dye mixture.

It is important that you never touch the TLC plates except when carefully holding them by the side.

1. Place about 8 mL of the solvent in the development chamber. Cover the chamber immediately.
2. Obtain a TLC plate from your instructor. Make sure you hold it by its side.
3. Place the plate on a piece of glass. Use a capillary tube to apply a small dot of the dye mixture to the plate, about 1/4 inch from the bottom of the plate.
4. Allow this dot to dry for about one minute.
5. Apply another small dot on top of the dried dot.
6. Consult with your instructor whether a third application is necessary.
7. Place the development chamber in a location where it will stand undisturbed for the next step.
8. Once the last application has dried, place the plate in the development chamber with the spotted end down, making sure that the solvent is not disturbed and does not splash on the spot. The solvent must be able to rise to the spot as it travels up

the plate. Make sure that the chamber is never moved once the plate has been placed inside, so that the solvent rises up smoothly and undisturbed. The TLC setup is shown in Figure 3.1.

9. Remove the plate from the chamber when the solvent is about 1 cm (½ inch) below the upper end of the plate. Use tweezers to remove the plate without disturbing the solvent at the bottom of the chamber.

Figure 3.1 TLC setup

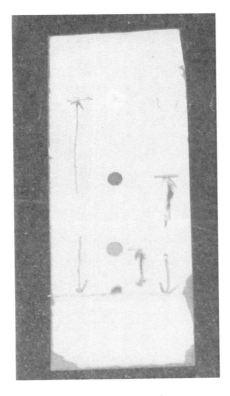

Figure 3.2 Thin layer chromatogram

22 Experiment 3 Chromatography

10. As soon as the plate has been removed, use a pencil to lightly mark the level where the solvent has stopped (upper end of the plate) as well as the initial point of spotting. This should be done immediately, because the solvent will quickly dry and the line between the wet and dry portions of the plate will be hard to identify.
11. Place the plate on the watch glass and let it dry. There should be two (or more) spots at some distance between the two pencil lines on the plate.
12. Circle each spot lightly and mark its center.
13. For each dye:
    i) Record its color on the data sheet
    ii) Measure (in mm) and record on the data sheet the distance ($d_\#$) this dye has traveled from the original point of application to the center of the spot
14. Measure (in mm) and record on the data sheet the distance (d_s) the solvent has traveled, as measured from the original point of application to the top pencil line.
15. Calculate the R_f value for each spot on the chromatographic plate, as seen in Figure 3.2.

Column Chromatography

Special items: Disposable pipet (column), cotton, metal rod, powdered stationary phase, solvents, and dye mixture.

1. Use a thin metal rod to insert a tiny piece of cotton to the bottom of the pipet.
2. Clamp the pipet to a ring stand, using a cork and a buret clamp as demonstrated by your instructor.
3. Fill the pipet with the indicated stationary phase, all the way up to about 1 inch from the top (Figure 3.3).

Figure 3.3 Column chromatography setup

4. Place a test tube under the pipet (column).
5. Use a dropper to fill the pipet with the solvent and tap the suspension with a metal rod to eliminate any trapped bubbles.
6. Add 3 drops of the methylene blue/fluorescein dye mixture on top of the stationary phase in the column. Make sure that the solvent **always** covers the column during the entire experiment to prevent drying and cracking of the stationary phase in the column.
7. Collect the solution of the first dye until it has completely eluted. Record its color on the data sheet.
8. Replace the test tube to collect the second dye. Record its color on the data sheet.

Title: Chromatography

Objective: To seperate Dye Mixture

Procedure:

Data:

Results:

Conclusion:

*Follow this Order

Experiment 4

Formula of a Hydrated Salt

Introduction

Many salts, such as calcium chloride, $CaCl_2$, naturally trap water molecules in their crystalline structure when exposed to humidity. These water-containing salts are called hydrates, or hydrated salts. Thus:

$$CaCl_2 \text{ (solid)} + 2H_2O \text{ (g)} \rightarrow CaCl_2 \cdot 2H_2O \text{ (solid)}$$

calcium chloride (anhydrous)   water   calcium chloride dihydrate (hydrated)

In determining the molar mass of a hydrate, the mass of water must be included. Thus, for $CaCl_2$ the molar mass is $40 + 2(35.5) = 111$ g/mol and for $CaCl_2 \cdot 2H_2O$ the molar mass is $40 + 2(35.5) + 2(18) = 147$ g/mol.

Heating the hydrated salt will reverse the reaction and remove the water in the form of vapor. For example;

$$CaCl_2 \cdot 2H_2O \text{ (solid)} \rightarrow CaCl_2 \text{ (solid)} + 2H_2O \text{ (g)}$$

calcium chloride (hydrated)   calcium chloride (anhydrous)   water

As a result, heating of a known quantity of a hydrated salt leads to the calculation of both the masses of the anhydrous salt and water and, consequently, the calculation of the relative ratio of moles of water per mole of the anhydrous salt in the formula.

In this experiment, a known mass of a hydrated salt will be heated to constant weight until all water is removed. Calculation of the relative mass of the anhydrous salt and water will lead to the relative molar ratio in the formula of the hydrated salt.

Safety Tips

Review from the first class:

1. Wearing goggles
2. Use of Bunsen burners or hot plates
3. Handling powdered chemicals

Follow your professor's instructions for the use and disposal of powdered chemicals.

Procedure

1. Calculate and record the molar mass for each anhydrous salt in Table 4.1.
2. Obtain an unknown hydrated salt and record its number on the data sheet (item A).
3. Obtain from your instructor the name of your anhydrous salt and find its formula in Table 4.1. Record the anhydrous formula in the data sheet (item K).
4. Set up a Bunsen burner, iron ring, ring stand, evaporating dish, and watch glass cover or crucible and cover as seen in Figure 4.1. Alternatively, you may use a hot plate.
5. Heat the evaporating dish or crucible (with the cover) under strong heat for 5 minutes. Allow to cool for about 5 minutes. Use of the evaporating dish usually yields better results.
6. Use tongs to transfer the evaporating dish (with the cover) to the balance and record its mass on the data sheet (item B).
7. While the evaporating dish is still on the balance, place 0.6–1.0 g of the hydrated salt in it and reweigh accurately. Record the **exact** mass on the data sheet (item C). You should record the actual mass. **Do not tare the evaporating dish!!!**
8. Use the tongs to place the evaporating dish on the iron ring of the setup in Figure 4.1.
9. Heat the evaporating dish for about 5 minutes.
10. Let the evaporating dish cool for about 5 minutes while still on the ring. Use tongs to weigh the cooled dish and record its mass on the data sheet (item D).
11. Transfer the evaporating dish on the iron ring again and reheat it for another 2 minutes and record its mass on the data sheet (item D). The two masses from steps 10 and 11 should match. If they do not match, reheat the evaporating dish for 2 more minutes and cool for 2 more.
12. Perform your calculations on the data sheet (items E through N) and provide your conclusion (item O).

100 mL Beaker
Hot Plate

Figure 4.1 Experimental setup for the dehydration of a salt

Experiment 5

Chemical Properties of a Substance: Chemical Reactions

Introduction

Chemical reaction, chemical change, or **chemical process,** is a special interaction between substances, called **reactants,** that leads to new substances, called **products.** Chemical reactions occur constantly in every aspect of nature, including the human body.

A chemical reaction is usually written in the form of a chemical equation when reactants and products are represented by their chemical formulas.

Often a chemical reaction is accompanied by a detectable change such as:

a) The evolution of gas
b) The formation of solid (precipitate)
c) The change in color
d) Loss of heat by the environment (cooling of the environment—**endothermic reaction**) or gain of heat by the environment (heating of the environment—**exothermic reaction**).

Note that a **physical process,** such as dissolution of various solids in water, can be also either exothermic or endothermic.

The chemical reactions that are accompanied by detectable changes are used in **qualitative analysis.** This is the process used to identify the presence of various substances, such as contaminants in water, air, and soil or to determine the purity of various products, such as medication in the pharmaceutical industry.

There are many quantitative aspects involved in the study of chemical reactions. For example, calculations are performed to determine how much product can be obtained from available amounts of reactants or how much heat will be absorbed by the environment from available quantities of reactants of a certain exothermic reaction.

Following are a few examples of chemical reactions:

1. A match is stricken near hydrogen gas, which will react with oxygen gas (from air) to produce steam (gaseous water):

$$2H_2(g) + O_2(g) \rightarrow 2H_2O(g)$$

2. Human and animal respiration is a reaction between glucose, which is found in their bodies, and oxygen from air to yield gaseous carbon dioxide and liquid water:

$$C_6H_{12}O_6(aq) + 6O_2(g) \rightarrow 6CO_2(g) + 6H_2O(l)$$

3. Addition of solid baking soda (NaHCO₃, sodium bicarbonate) to the stomach juices (HCl, hydrochloric acid) yields table salt (NaCl, sodium chloride), gaseous carbon dioxide, and liquid water:

$$NaHCO_3(s) + HCl(aq) \rightarrow NaCl(aq) + CO_2(g) + H_2O(l)$$

Safety Tips

Review from the first class:

1. Wearing goggles
2. Handling solid, liquid, and gaseous chemicals
3. Handling glassware in which mixing of chemicals occurs

Follow your professor's instructions on the disposal of leftover chemicals of your experiments.

Procedure

I. Demonstration by Instructor

1. A strip of magnesium is placed in a crucible or evaporating dish and burned. On the data sheet, record your observations, such as the appearance of magnesium before and after burning. Provide a balanced chemical equation for the chemical reaction that takes place.
2. 0.5 g of solid barium hydroxide and 0.5 g solid ammonium thiocyanate are placed into a test tube and mixed with a glass rod. Record your observations such as the appearance of each reactant and product and whether you have an exothermic or endothermic reaction. Provide a chemical equation for the reaction that takes place.

II. Performance by Students

For each of the chemical reactions (A through G):

1. Observe and record on the data sheet the appearance of every reactant: if pure, describe its physical state (solid or liquid) and color; if in solution, describe its color. For instance: Zn(s) is a gray solid, CuSO₄(aq) is a blue solution, etc.
2. Place each pair of reactants in separate test tubes and label each test tube A through G.
3. Observe and record all visual changes (appearance of every product) on the data sheet. Changes may include color change, change of physical state (evolution of gas), solubility change (appearance of solid, precipitate), or change in temperature (colder or warmer), which corresponds to an endothermic or exothermic reaction.
4. Provide an equation for each of the chemical reactions that take place.

Experiment 5 Chemical Properties of a Substance: Chemical Reactions **35**

Chemical Reactions

A. Place a small amount of mossy zinc (equal to covering the tip of a spatula) in a small test tube. Add about 10 drops of copper (II) sulfate solution. Allow the reaction to take place and record your results at the end of the laboratory period.
B. Place a strip of magnesium in a test tube. Slowly and carefully add 5–10 drops of 1 M hydrochloric acid solution. Record your observations as indicated above.
C. Place 10 drops of 1 M sodium chloride solution and 10 drops of 1 M silver nitrate solution in a test tube. Record your observations on the data sheet.
D. Place 10 drops of 1 M barium chloride solution and about 10 drops of 1 M sodium sulfate solution in a test tube. Record your observation on the data sheet.
E. Place 10 drops of 1 M lead (II) nitrate solution and 10 drops of 1 M potassium chromate solution in a test tube. Record your observation on the data sheet.
F. Place 10 drops of 1 M ferric chloride solution and 10 drops of 1 M potassium thiocyanide solution in a test tube. Record your observations on the data sheet.
G. Place 3–4 crystals of solid sodium bicarbonate in a test tube. Slowly add about five drops of 1 M hydrochloric acid solution. Record your observations on the data sheet.

5. Do not forget to record your observations for step A.
6. *Dissolution of ammonium chloride in water:* Place a few crystals of ammonium chloride in a small test-tube. Add about 5 mL of water and shake the contents to dissolve the ammonium chloride. Observe the temperature change. Record your observations on item H on the data sheet.
7. *Dissolution of sodium bicarbonate in water:* Repeat step 5 using sodium bicarbonate instead of ammonium chloride. Record your observations in item I on the data sheet.
8. Students perform computer simulation experiment "Calorimetry: Developing a New Ice Pack" using "Bringing to the Lab" software to facilitate and assess their understanding of exothermic and endothermic processes; to learn quantitative aspects associated with chemical and physical processes; and to assist in lecture calculations where concepts of specific heat and enthalpy are used. The results are printed and attached to the report.

1) Combination Synthesis Reaction: $A + B \rightarrow AB$

2) Decomposition: $AB \rightarrow A + B$
   $H_2O \rightarrow H_2 + O_2$

3) Single Displacement: $A + BC \rightarrow AB + C$
   $Pb + NaCl \rightarrow PbCl_2 + Na$

4) Double Replacement: $AB + CD \rightarrow AC + BD$

Experiment 6

Electrical Conductivity of Aqueous Solutions: Electrolytes and Nonelectrolytes

Introduction

Ionic compounds are composed of positive and negative **ions** arranged in characteristic patterns. They are usually solids at room temperature, which indicates that their ions are "frozen" in a crystalline form. Ionic compounds dissociate into **cations** (positively charged ions) and **anions** (negatively charged species) when dissolved in water. Each ion becomes surrounded by water molecules and is represented with the shorthand notation (aq), for aqueous. Thus, ionic solids (s) dissociate as follows in water:

$$NaCl\,(s) \xrightarrow{H_2O} Na^+(aq) + Cl^-(aq)$$

$$MgBr_2\,(s) \xrightarrow{H_2O} Mg^{2+}(aq) + 2Br^-(aq)$$

As a result of their dissociation in water, aqueous ions become mobile (as compared to being stationary in their solid state) and are able to conduct electric current.

Similarly, some polar molecular compounds also dissolve in water and may ionize into cations and anions. Some of these compounds ionize almost completely, while others do so only partially. Ionization of these molecular compounds in water can be represented by chemical equations:

$$HCl\,(aq) \rightarrow H^+(aq) + Cl^-(aq),$$

$$HNO_3\,(aq) \rightarrow H^+(aq) + NO_3^-(aq)$$

Compounds, such as the ones described previously, that dissociate (ionize) almost exclusively into ions when in aqueous solutions are called **strong** electrolytes and easily conduct electric current.

A large number of polar molecular compounds dissolve in water with practically no ionization. In this case, the molecules of the compound are separated from each other and are surrounded by the large excess of water molecules. Since there are no mobile ions available, these compounds do not conduct electricity when in solution and are called nonelectrolytes. An example of such a compound is table sugar, $C_{12}H_{22}O_{11}$:

$$C_{12}H_{22}O_{11}(s) \xrightarrow{H_2O} C_{12}H_{22}O_{11}(aq) \text{ (no ions)}$$

The behavior of most polar molecular compounds lies between the strong and nonelectrolyte groups. These compounds, which ionize only partially when dissolved in water and conduct only a weak electric current, are called weak electrolytes. Examples of such compounds are acetic acid, $HC_2H_3O_2$, and ammonia, NH_3. Acetic acid ionizes as follows:

$$HC_2H_3O_2 \text{ (aq)} \leftrightarrows H^+ \text{ (aq)} + C_2H_3O_2^- \text{ (aq)},$$

where the double arrow indicates that some ions are surrounded by water shown on the right, and some molecules of acetic acid are surrounded by water shown on the left of the equation. Similarly, ammonia's ionization is given by the following equation:

$$NH_3 \text{ (aq)} + H_2O \text{ (l)} \leftrightarrows NH_4^+ \text{ (aq)} + OH^- \text{ (aq)}$$

In this experiment the conductivity of various substances and solutions will be tested via the conductivity apparatus seen in Figure 6.1. Its electrodes will be dipped in the various substances and solutions. Strong electrolytes will produce a bright light while poor electrolytes will yield a dim light. No light should be detected in the case of nonelectrolytes.

Safety Tips

Review from the first class:

1. Wearing goggles
2. Handling liquid and solid chemicals

Follow your professor's directions about the use of the light bulb assembly (conductivity apparatus).

Procedure

1. Set up the conductivity apparatus (light bulb assembly) as seen in Figure 6.1.
2. Make sure that the switch on the light bulb wire is off. **Do not touch the electrodes when the assembly is plugged in, even if the switch is off!!!**
3. Connect the assembly to the electric outlet.
4. Record the name, chemical formula, class (ionic or covalent, acid, base, or salt), and physical state (solid or liquid) of each pure compound on the data sheet (Table 6.1).
5. Test the electric conductivity of each pure compound in Table 6.1 by doing the following:
    a) Fill a 50 mL beaker halfway, with the compound to be tested.
    b) Immerse the electrodes of the light bulb assembly into the compound (or solution) in the beaker.
    c) Turn on the switch of the light bulb.
    d) Record your observation (bright light, dim light, or no light) and conclusion whether the compound conducts electricity.

Experiment 6 Electrical Conductivity of Aqueous Solutions: Electrolytes and Nonelectrolytes **43**

Figure 6.1 Light bulb (conductivity) apparatus setup

6. Classify each compound as strong electrolyte, weak electrolyte, or nonelectrolyte based on the conductivity behavior of its aqueous solution.
7. Repeat steps 4 through 7 for the various solutions in Table 6.2.
8. Students perform computer simulation of the Qualitative Analysis using "Bridge to the Lab" software—to facilitate and assess their understanding of the chemical reactions accompanied by detectable changes. The procedure appears in Experiment 13.

Questions and Problems

(Show all calculations)

1. Name each of the ionic compounds below and provide the equation representing its complete dissociation in water:

   a) $NaOH \rightarrow$ Na(aq) + OH(aq) – Sodium hydroxide

   b) $Ca(OH)_2 \rightarrow$ Ca^{+2} + 2OH – Calcium hydroxide

   c) $KCN\ (s) \rightarrow$

   d) $Na_2CO_3 \rightarrow$

   e) $K_3PO_4 \rightarrow$

   f) $Ba(NO_3)_2 \rightarrow$

   g) $Mg(NO_2)_2 \rightarrow$

   h) $NH_4Cl \rightarrow$

   i) $(NH_4)_2SO_4 \rightarrow$

2. For each of the acids below provide the equation representing its ionization in water, name each compound:

   a) $HCl \rightarrow$

   b) $HNO_3 \rightarrow$

   c) $HClO_4 \rightarrow$

   d) $H_2SO_4 \rightarrow$

Experiment 7

Molar Mass of a Volatile Gas

Introduction

Molar mass of a substance is defined as the mass of one mole of this substance. If an arbitrary sample of a substance is available and its mass (m) and number of moles (n) are known, then the molar mass or molecular weight (MW) is calculated as the ratio of mass to moles:

$$MW = m/n \quad \text{(eq. 1)}$$

where:

$$m = \text{mass (in g) and}$$
$$n = \text{\# of moles}$$

If a known sample of a volatile liquid is available, its mass (m) can be measured. That liquid can then be vaporized by heating. Consequently the pressure, volume, and temperature of the gas may be measured. As a result the number of moles in the sample can be calculated using the Ideal Gas Law:

$$PV = nRT \quad \text{(eq. 2)}$$

where

P is pressure (in atm)
V is volume (in L)
n is number of moles
R is the gas constant (0.0821 atm·L/mol·K) and
T is temperature of the gas (in K).

Finally the molar mass (or molecular weight) (MW) can be calculated using equation (1).

Safety Tips

Review from the first class:

1. Wearing goggles
2. Handling, heating, and disposal of liquid chemicals
3. Use of Bunsen burner or hot plate

Procedure

1. Weigh a clean and dry (do not wash!) boiling (Florence) flask together with its aluminum foil cover. Record the mass on the data sheet (item A).
2. Place 3–4 mL of the unknown volatile liquid in the flask and immediately cover tightly with the aluminum foil.
3. Record the unknown number (or letter designation) on the data sheet (item I).
4. Use a pin to make a tiny hole in the aluminum foil cover.
5. Assemble the setup for heating the flask in a water bath as shown in Figure 7.1.
6. Fill the beaker (water bath) with water to the level that will keep the greater part of the flask immersed.
7. Heat the water bath until the volatile liquid in the flask has vaporized (about 5 minutes **after** the water has started to boil).
8. Record the temperature of the water bath on the data sheet (item B). This is also the temperature of the gas sample in the flask.
9. Remove the flask from the water bath and allow it to cool to room temperature. You should be able to see the vapor condensing and droplets accumulating inside the flask.
10. Use a paper towel to gently wipe any water droplets on the outside of the flask and on the aluminum foil cover.

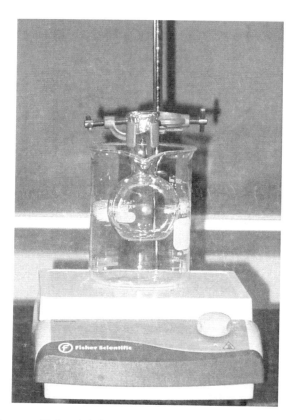

Figure 7.1 Molar mass of a volatile liquid setup

11. Weigh the flask, aluminum foil, and the contents of the flask. Record the mass on the data sheet (item C).
12. Dispose of the volatile liquid condensate from the flask as directed.
13. Fill the flask to the very top with water. Measure the volume of water by pouring it into a graduated cylinder. You will have to use the graduated cylinder repeatedly to measure the total volume of water.
14. Record the volume on the data sheet (item D). This is equal to the volume of the flask and the volume of the sample when it was in the gaseous state.
15. Record the atmospheric (barometric) pressure on the data sheet) (item E).
16. Calculate the molar mass of the unknown liquid.

- Graduated Cylinder
- Florence Flask
- Funnel
- Pin
- Aluminum foil
- Unknown liquid
- hot plate
- Ring stand
- beaker

Experiment 8   Determination of a Solution's Concentration by Visible Spectrophotometry   59

8. Replace the blank with a cuvette filled with the stock solution (from step 4). Make sure to wipe the cuvette before inserting it into spectrophotometer.
9. Record the absorbance of the stock solution in Table 8.1 of the Data sheet and remove the cuvette filled with the stock solution from the spectrophotometer.
10. Reset the wavelength, increasing it by an increment of 5 nm (to 355 nm).
11. Repeat steps 6–9.
12. Repeat steps 10–11 until the final wavelength of 750 nm is recorded.
13. Select the wavelength at which the highest absorbance occurred after consultation of the data you recorded in Table 8.1. This is the wavelength that will be used in Part II of this experiment.

II: Determination of the Unknown Concentration of a Solution Using Beer's Law

The first step is constructing a **calibration curve** (plot of absorbance vs concentration) using a series of known solutions of the same compound. These solutions will be made by the students who will prepare various dilutions of the stock solution. The calibration curve gives access to a large number of concentrations and their corresponding absorbances. The absorbance value of any unknown solution is then obtained via the spectrophotometer and is marked on the absorbance y-axis of the calibration curve plot. The value for the concentration of the unknown solution is found on the concentration x-axis.

1. Use a ring stand to clamp one 25 mL and one 50 mL burettes.
2. **Use a funnel to carefully fill the 25 mL burette** with the stock solution ("A") of potassium dichromate (K_2CrO_4) to the 0.0 mL level.
3. Use a funnel to fill in the 50-mL burette with distilled water to the 0.0 mL level.
4. Obtain three beakers, 50 mL, 100 mL, and 150 mL. Label them B, C, and D, respectively.
5. Dispense 5.00 mL of stock solution from the 25 mL buret into each of the beakers: B, C, and D.
6. Dispense the following volumes of distilled water from the 50 mL buret:
   - 5.0 mL into beaker B (total volume = 5.0 mL + 5.0 mL = 10.0 mL),
   - 15.0 mL into beaker C (total volume = 5.0 mL + 15.0 mL = 20.0 mL),
   - 35.0 mL into beaker D (total volume = 5.0 mL + 35.0 mL = 40.0 mL).
7. Mix throughly the solutions from step 6.
8. Obtain an unknown solution ("X") from your instructor and record its number (if any) on the data sheet.
9. Place a small test tube (cuvette) in the mini test tube rack from Part I.
10. Fill the cuvette up to the 3/4 level with distilled water.
11. Set the spectrophotometer to the wavelength at which the maximum absorbance occurs (determined in Part I).
12. Using cuvette with distilled water as the blank ("E"), set the spectrophotometer to 0 absorbance as done previously in steps 6 and 7 of part I.
13. Refill the cuvette with solution D and record its absorbance on the data sheet in Table 8.2.
14. Repeat step 13 for solutions C, B, A, and X.

15. Construct a calibration "curve"—a plot of absorbance vs concentration using solutions A, B, C, D, and E. The absorbance values for A, B, C, D, and E are on the y-axis and corresponding concentrations of A, B, C, D, and E are on the x-axis. Draw the best fit straight line that intercepts the origin.
16. Mark the value for absorbance of solution X on the y-axis.
17. Draw a horizontal line from absorbance of X on the y-axis until it intersects the calibration curve.
18. Draw a vertical line from the previous intersection to the x-axis—this is the concentration of unknown solution X. Record it on the data sheet in Table 8.2 and next to the unknown number.

Questions and Problems

(Show all calculations)

1. Convert 370 nm to
    a) meters (m)

    b) centimeters (cm)

    c) millimeters (mm)

2. For the light, whose wavelength is 370 nm, identify:
    a) The color range: _____

    b) The color range of the complementary: _____

    c) The nanometer (nm) range of the complementary: _____

3. For the light, which is in the green color range, identify:
    a) The nanometer (nm) range of this wavelength: _____

    b) The complementary color range: _____

    c) The nanometer (nm) range of the complementary: _____

    d) A green compound absorbs _____ color range.

4. You have two solutions of the same compound that are blue. One is very pale and the other is very intense. Absorbance by each solution was measured in a spectrophotometer at the same maximum wavelength at which the solution absorbs.
    a) Which solution has the higher absorbance?

    b) Which is the more concentrated solution?

5. You have two solutions of the same pink color. The concentration of the first solution is 1.0 M, while the concentration of the other solution is 3.0 M. Absorbance of each solution was measured in the same spectrophotometer at the max wavelength at which the solution absorbs.
    a) Which solution has the more intense pink color?

    b) Which solution has the higher absorbance?

Experiment 9

pH Determination of Solutions

Introduction

Acids (general formula HA) are polar molecular compounds that are proton donors in solution. Thus:

$$HA\ (aq) \rightarrow H^+\ (aq) + A^-\ (aq),$$

where

(aq) indicates an aqueous (or water) solution.

Strong acids ionize completely, while **weak acids ionize only to some degree.** For instance, HCl (strong acid) and $HC_2H_3O_2$ (weak acid) ionize as follows:

$$HCl\ (aq) \rightarrow H^+\ (aq) + Cl^-\ (aq)$$
$$HC_2H_3O_2\ (aq) \leftrightarrows H^+\ (aq) + C_2H_3O_2^-\ (aq)$$

A single arrow from left to right indicates that the ionization is almost 100%. The double arrow indicates that the weak acid exists in solution both as unionized molecules as well as ions.

Most acids are **monoprotic** because they release only one proton (H^+) when ionized. **Diprotic** acids release two protons, while **triprotic** acids release three protons. Thus:

$$HCl\ (aq) \rightarrow H^+\ (aq) + Cl^-\ (aq)$$

$$H_2SO_4\ (aq) \rightarrow 2H^+\ (aq) + SO_4^{2-}\ (aq)$$

$$H_3PO_4\ (aq) \rightarrow 3H^+\ (aq) + PO_4^{3-}\ (aq)$$

The concentration of an acid in a solution is expressed in **molarity** units (moles of solute/liter of solution). For strong monoprotic acids, such as HCl and HNO_3, the concentrations of H^+ and the negative ion (also called **counter ion**, in this case Cl^- and NO_3^- respectively) are equal to each other and each is equal to the initial concentration of the acid. For weak acids, however, the proton concentration cannot be as easily related to the concentration of the acid.

The proton concentration of acids may be expressed in either a) **decimal** form. In this case, 0.1 M solution of HCl (strong acid) has

$$[H^+] = 0.1\ M = 1 \times 10^{-1}\ M\ \text{and}$$

0.001 M solution of HNO_3 (also strong acid) has

$$[H^+] = 0.001 \text{ M} = 1 \times 10^{-3} \text{ M, or}$$

b) **pH scale.** In this situation,

$$pH = -\log [H^+].$$

Thus, a 0.01 M HCl solution has

$$[H^+] = 0.01 \text{ and } pH = -\log [H^+] = +2.$$

Similarly, a 0.0001 M HNO_3 solution has

$$[H^+] = 0.0001 \text{ M and } pH = -\log [H^+] = +4.$$

Note that as the proton concentration decreases (i.e., acidity of the solution decreases), the pH of the solution increases. Conversely, as the proton concentration increases (i.e., acidity of solution increases), the pH of the solution decreases.

Bases are substances that generate hydroxide ions (OH^-) in solution. When dissolved in water, metal hydroxides (ionic compounds) dissociate 100% into positive ions of the metals and negative hydroxide ions (OH^-). This makes metal hydroxides to be strong bases. For example:

$$NaOH \text{ (s)} \xrightarrow{H_2O} Na^+ \text{ (aq)} + OH^- \text{ (aq)}$$

The left-hand side of the equation refers to NaOH (s = solid) before it was dissolved in water. The right side of the equation shows dissociation of NaOH into positive (cations) and negative (anions) ions.

Polar molecular compounds that have at least one lone pair of electrons available on a very electronegative atom, such as nitrogen in ammonia (NH_3) for example, are classified as weak bases. Ammonia ionizes in water as follows:

$$NH_3 \text{ (aq)} + H_2O \text{ (l)} \leftrightarrows NH_4^+ \text{ (aq)} + OH^- \text{ (aq)}$$

Weak bases ionize, in a fashion similar to weak acids, only to some degree. The degree of ionization is different and characteristic for each compound.

The concentration of a base in a solution is also reported as moles of the metal hydroxide per one liter of the solution (also known as molarity units). Bases, such as NaOH or KOH, have concentration of OH^- equal to the original concentration of NaOH or KOH before ionization. On the other hand, one mole of $Ca(OH)_2$ generates two moles of OH^-, since there are two hydroxide ions per one formula unit of the compound. The dissociation of this compound will be:

$$Ca(OH)_2 \text{ (s)} \xrightarrow{H_2O} Ca^{2+} \text{ (aq)} + 2OH^- \text{ (aq)}$$

The power of 10 scale can also be used to measure the concentrations of OH^- in solution in a fashion similar to pH. It is known as the **pOH scale**. Thus, 0.1 M solution of NaOH (strong base) has:

$$[OH^-] = 0.1 \text{ M} = 1 \times 10^{-1} \text{ M} \quad \text{and} \quad pOH = -\log [OH^-] = +1.$$

Similarly, a 0.001 M solution of KOH (strong base) has:

$$[OH^-] = 0.001 \text{ M and pOH} = -\log [OH^-] = +3.$$

The lower the OH^- concentration, the higher the pOH of a solution (i.e., the lower the basicity of the solution). The higher the OH^- concentration, the lower the pOH of a solution (i.e., the higher the basicity of the solution).

Water behaves both as a weak acid and a weak base. This is expressed by the equation:

$$\underset{\text{acid}}{H_2O \text{ (l)}} + \underset{\text{base}}{H_2O \text{ (l)}} \leftarrow H_3O^+ \text{ (aq)} + OH^- \text{ (aq)}$$

This equation is referred to as the **autoionization** of water. For pure water the concentrations of $[H^+] = [H_3O^+]$ and $[OH^-]$ are identical and equal to 1×10^{-7} M. Thus:

$$pH = pOH = 7 \quad \text{and} \quad pH + pOH = 14 \text{ for pure water.}$$

When an acid is dissolved in water the H^+(aq) concentration increases and the OH^-(aq) concentration decreases compared to pure water. For example, an acidic solution that has:

$$[H^+] = 1 \times 10^{-5} \text{ M}, \quad \text{has pH} = 5, \quad \text{consequently}$$

$$pOH = 14 - 5 = 9, \quad \text{and} \quad [OH^-] = 1 \times 10^{-9} \text{ M}.$$

Similarly, a basic solution that has $[OH^-] = 1 \times 10^{-5}$ M, has pOH = 5. As a result:

$$pH = 14 - 5 = 9 \quad \text{and} \quad [H^+] = 1 \times 10^{-9} \text{M}.$$

Clearly, an acidic solution will have a pH lower than 7 and pOH higher than 7, while a basic solution will have a pOH lower than 7 and a pH higher than 7.

When a compound other than an acid or base is dissolved in water (e.g., table salt or sugar), neither the pH nor the pOH changes compared to pure water, which corresponds to a neutral solution (i.e., pH = 7 and pOH = 7).

The instrument that measures the proton concentration of a solution and expresses it using the pH scale is called the **pH meter.** The part of the instrument that performs the measurement is a sensor—also known as the **pH probe.**

There are also compounds that are used as **indicators** for the pH estimation of a solution. An indicator is usually a large organic molecule, which changes its color with pH change.

Salts are ionic compounds that dissociate when in solution but yield neither H^+ nor OH^-. For example, sodium chloride (table salt) dissociates as follows:

$$NaCl(s) \xrightarrow{H_2O} Na^+(aq) + Cl^-(aq)$$

A **salt** can be considered as the **product** of the reaction of an acid and a base. Thus, NaCl is the product of the reaction of HCl and NaOH:

$$\underset{\text{acid}}{HCl \text{ (aq)}} + \underset{\text{base}}{NaOH \text{ (aq)}} \rightarrow \underset{\text{salt}}{NaCl \text{ (aq)}} + \underset{\text{water}}{H_2O \text{ (l)}}$$

68 Experiment 9 pH Determination of Solutions

> ## Safety Tips
> Review from the first class:
> 1. Wearing goggles
> 2. Handling liquid and solid chemicals as well as solutions
>
> **Special case:** Handling solutions of acids and bases.
> Follow all instructions given by your professor in the operation and handling of the pH meter.

Procedure

NOTE: Make sure that all transferring of the solutions should be made using funnels.

Dilutions of the 1 M acidic stock solution:

1. Place about 100 mL of the 1 M acidic stock solution in a 250 mL beaker and label this beaker with the integer "0."
2. Use a 10 mL graduated cylinder to transfer 10 mL of this solution into another 250 mL beaker and label this new beaker with the integer "1." Add 90 mL of distilled water using a 100 mL graduated cylinder and stir.
3. Use a third 250 mL beaker to repeat step 2 placing 10 mL of the solution in the beaker labeled with the integer "1," adding 90 mL of distilled water and stirring. Label this beaker with the integer "2."
4. Continue dilutions as described above until you prepare a solution in a beaker labeled with the integer "6."

Dilutions of the 1 M Basic Stock Solution

1. Place about 100 mL of the 1 M basic stock solution in a 250 mL beaker and label this beaker with the integer "14."
2. Use a 10 mL graduated cylinder to transfer 10 mL of this solution into another 250 mL beaker and label this new beaker with the integer "13." Add 90 mL of distilled water using a 100 mL graduated cylinder and stir.
3. Use a third 250 mL beaker to repeat step 2, placing 10 mL of this solution in the beaker labeled with the integer "1," adding 90 mL of water and stirring. Label this beaker with the integer "12."
4. Continue dilutions as described until you prepare a solution in a beaker labeled with the integer "8."

Measuring pH of the Solutions

1. Place 100 mL of distilled water in a 250 mL beaker and label it with the integer "7."
2. Measure the pH of the distilled water (beaker with integer "7") using the pH meter and record it on the data sheet.

3. Measure the pH of each acidic solution (starting with the beaker labeled with the integer "6" and continuing all the way to beaker labeled "0") using the pH meter and record each pH in the data table.
4. Measure the pH of each basic solution (starting with the beaker labeled with the integer "8" and continuing all the way to beaker labeled "14") using the pH meter and record each pH in the data table.

Use of Indicators

NOTE: In addition to the recording of color changes produced by the universal indicator in the procedure that follows, use the color charts to estimate and record the pH on the data sheet.

1. Place 14 test tubes in a test tube rack and label each test one with the integers from 0 to 14.
2. Fill in (to half the test tube volume) each of the test tubes with the corresponding solutions from beakers labeled 0 to 14. (Remember that number 7 is distilled water.)
3. Select an indicator and record its color in the data table.
4. Add 2–3 drops of the indicator to each test tube, using a medicine dropper or a disposable pipette. Swirl gently the contents of each test tube.
5. Record the color of the solutions in each test tube on the data sheet.
6. Discard the solutions and wash the test tubes thoroughly.
7. Repeat steps 2–6 using a different indicator until all indicator solutions have been tested.
8. Wash the test tubes again and repeat step 2.
9. Dip a clean stirring rod in the first test tube and touch a strip of dry (paper) indicator that has been placed on a dry watch glass.
10. Record the color of the strip on the data sheet.
11. Discard the strip of paper in the paper trash (**NOT in the sink**) and rinse the watch glass with water.
12. Repeat steps 9–11 for the next test tube using a new strip of the same indicator.
13. Repeat steps 9–12 using a different paper indicator.
14. Do not forget to record the original color of the paper indicator in the data sheet.

Indicators to be Used

1. Universal solution **and** paper
2. Methyl Red solution
3. Thymol Blue solution
4. Bromphenol blue solution
5. Bromthymol blue solution
6. Phenolphthalein solution
7. Litmus—**both** pink and blue paper

Computer Procedure for pH Determination of Solutions

1. Calibrate your pH sensor following the directions of the calibration procedure.
   a) Obtain a small amount of pH buffers and place them in separate small beakers. It is customary to use buffers in the range of pH= 4–9.
   b) On the interface unit itself, have your pH sensor connected to one BNC receptor. Place the sensor into the beaker that is holding the lowest pH buffer, making sure that the tip of the sensor is covered by the buffer solution.
   c) On the sensor calibration menu, click on "pH."
   d) The pH sensor window will then open. Look at the sensor voltage and wait until the mill volt reading stabilizes. Type into the space marked "enter the pH" the value of your buffer.
   e) Rinse the pH sensor with distilled water as you change solutions.
   f) Place the pH sensor into the highest pH buffer and enter its value.
   g) The sensor is now calibrated.
2. After calibration, set up the program to plot pH (on the y-axis) and keyboard entry (on the x-axis).
3. Obtain various samples in small beakers.
4. Start the program to collect the corresponding pH data.
5. The window will open asking for the entry place of the pH sensor into one of the pH samples. Once the reading stabilizes enter the solution number or name.
6. Repeat the previous procedures with all your solutions.
7. Print your graph.

Questions and Problems

(Show all calculations)

1. Classify each of the compounds below as strong or weak acids, strong or weak bases, or salts:

   a. HCl _____

   b. KOH _____

   c. NH_4NO_3 _____

   d. H_2SO_4 _____

   e. NH_4OH _____

   f. H_3PO_4 _____

   g. NaOH _____

   h. $HClO_4$ _____

   i. $Mg(OH)_2$ _____

   j. H_2CO_3 _____

   k. $BaSO_4$ _____

   l. $HC_2H_3O_2$ _____

2. Provide equations for each of the following:
   (Use any of the compounds that appear in question 1, if needed)

   a. Dissociation of a strong base in water solution:_____

   b. Ionization of a strong acid in water solution:_____

   c. Ionization of a weak acid in water solution:_____

   d. Ionization of a weak base in water solution:_____

   e. Autoionization of water:_____

3. A 5.0 M solution of HCl in water has a proton concentration of _____ M.

4. A 0.01 M aqueous NaOH solution has a hydroxide concentration of _____ M.

5. An aqueous HCl solution has a proton concentration equal to 6.00 mol/L. The HCl concentration in this solution is _____ M.

6. An aqueous KOH solution has a hydroxide concentration equal to 0.02 mol/L. The KOH concentration in this solution is _____ M.

7. The pH of an aqueous solution is 4.
   a. The pOH of this solution is _____

   b. [H⁺] is _____ (scientific notation) = _____ (decimal notation)

   c. [OH⁻] is _____ (scientific notation)

   d. Is this solution acidic or basic?

8. The pH of a solution is 12.

   a. The pOH of this solution is _____

   b. [H⁺] is _____ (scientific notation)

   c. [OH⁻] is _____ (scientific notation) = _____ (decimal notation)

   d. Is this solution acidic or basic?

9. The pOH of a solution is 3.

   a. The pH of this solution is _____

   b. [H⁺] is _____ (scientific notation)

   c. [OH⁻] is _____ (scientific notation) = _____ (decimal notation)

   d. Is this solution acidic or basic?

Experiment 10

Determination of a Solution's Concentration by Titration

Introduction

Solutions of two different solutes when mixed together very often produce a chemical reaction between the solutes. The solvent (in this case, water) does not participate in the reaction, as it just provides the medium in which the reaction takes place. Often the reaction between two compounds can take place only if they are dissolved in water. In a metathesis, or double replacement, reaction, the cation (positive ion) of one reactant reacts with the anion (negative ion) of the second reactant to yield one of the products. At the same time, the anion of the first reactant will react with the cation of the second reactant to yield the second product. Thus;

$$HCl\ (aq) + KOH\ (aq) \rightarrow H_2O\ (l) + KCl\ (aq),$$

where (aq) indicates that all compounds, both reactants and products, are dissolved in water.

The reaction between an acid and a base (reactants) generally produces water and a salt. This is called a **neutralization** reaction. The actual chemical change involves only protons (which make the solution acidic) and hydroxide ions (which make the solution basic) yielding the net ionic equation:

$$H^+\ (aq) + OH^-\ (aq) \rightarrow H_2O\ (l)$$

Since potassium chloride is completely soluble in water, the net ionic equation does not include KCl but instead the ions K^+ (aq) and Cl^- (aq) are excluded.

Complete neutralization is achieved only if the acid and the base (or the protons and the hydroxide ions) are used in equally proportional quantities of moles, which are shown as the coefficients of the chemical equation. In the example of neutralization, between HCl and KOH, the reactants react in a 1:1 ratio of moles (coefficients of the previous equations are 1 for each reactant).

If the concentrations of both acidic and basic solutions are identical, then equal volumes of each lead to complete neutralization. If the concentrations are not equal, and the volume of one of the solutions is fixed, then the volume of the other solution that is required to achieve neutralization can be calculated. At the neutralization point the number of moles of the acid (n_a) is equal to the number of moles of the base (n_b):

$$n_a = n_b \quad \text{or}$$

$$M_a \times V_a = M_b \times V_b \quad \text{(equation 1)}$$

where

M_a and V_a are the molarity and volume of the acid and

M_b and V_b are the molarity and the volume of the base.

Recall the definition of molarity: M = moles/L solution. The laboratory procedure for the determination of an unknown concentration via neutralization involves the **titration** process.

In the neutralization procedure, the exact volume of one reactant is measured using a **pipette** and placed into an Erlenmeyer flask. The solution of the second reactant is then placed into a **burette,** usually to the top division (see Figure 10.1). The flask with the solution of the first reactant is placed under the burette and an **indicator** that clearly distinguishes between acidic and basic solutions via color change is added to the flask. Then, the solution from the burette is added dropwise to the solution in the flask via the titration procedure, until the indicator signals the completion of neutralization by changing its color. At this point the titration stops and the volume of the solution drawn from the burette is measured. Finally, equation (1) is used to calculate the molarity of the unknown solution for the chemical reactions that have 1:1 ratio of reactants.

In this experiment aqueous solutions of acetic acid (vinegar) of unknown concentration and sodium hydroxide (base) of known concentration are used. The neutralization reaction between them will be:

$$HC_2O_2H_3 \text{ (aq)} + NaOH \text{ (aq)} \rightarrow H_2O \text{ (l)} + NaC_2O_2H_3 \text{ (aq)}$$

The concentration of a vinegar solution will be determined by titration of this solution against a sodium hydroxide solution of known concentration.

Safety Tips

Review from the first class:

1. Wearing goggles
2. Handling liquid and solid chemicals
3. Handling solutions of chemicals
4. Special case: solutions of acids and bases

Follow any additional instructions given by your professor, such as the use of a burette and a pipette.

Procedure

1. Place about 30 mL of vinegar (acid) into a 100 mL beaker.
2. Use a pipette to transfer exactly 10.0 mL of vinegar (from step 1) into each of the two 125 mL Erlenmeyer flasks, labeled as #1 and #2.
3. Record 10.0 mL as the volume of acid used (V_a) on the data sheet for trials 1 and 2.
4. Add about 10 mL of distilled water to each of the flasks.
5. Add about two drops of phenolphthalein to each of the flasks.
6. Clamp a burette to a ring stand as shown in Figure 10.1.
7. **Use a funnel** to carefully fill the burette with an aqueous 1.0 M NaOH solution to the zero level. Add the NaOH solution slowly to avoid overflowing the burette. This is the initial volume of base. Record it as V_i on the data sheet for trial 1.
8. Also record 1.0 M as the molarity of the base (M_b) on the data sheet for each trial.
9. Place Erlenmeyer flask #1 under the burette and a white sheet of paper under the flask.
10. Titrate dropwise the contents of the flask with the NaOH solution from the burette, with constant swirling. Stop titrating as soon as the solution turns barely pink. Note that any further addition of NaOH will have very little further effect on the color of the solution and it will stay basically unchanged. **Your goal is to stop when the solution just changed from clear to pink.**
11. Read the volume of the burette. This is the final volume of the base. Record it as V_f on the data sheet for trial 1.
12. Repeat steps 9–11 for Erlenmeyer flask #2, and record the initial and final volumes in the data table for trial 2. Note that V_i for the second trial is not going to necessarily be the same as V_i for the first trial.
13. Repeat steps 9–11 for Erlenmeyer flask #3.
14. Follow the directions on the data sheet to calculate the percent by mass-to-volume content of acetic acid in vinegar.

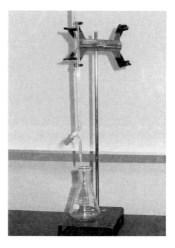

Figure 10.1 Titration setup

Measurements
A:
B: 0, 0, 0,

Experiment 11

Titration of Buffers

Introduction

Recall that strong acids, such as HCl, ionize entirely in aqueous solutions to yield protons and the chloride anion:

$$HCl\ (aq) \rightarrow H^+\ (aq) + Cl^-\ (aq)$$

In contrast, weak acids, such as HClO, ionize only partially in aqueous solutions:

$$HClO\ (aq) \rightarrow H^+\ (aq) + ClO^-\ (aq)$$

The equation is correctly written using water as a reactant in the following way:

$$HClO\ (aq) + H_2O\ (l) \rightarrow H_3O^+\ (aq) + ClO^-\ (aq)$$

Also recall that the reactants of a neutralization reaction are an acid and a base, and the products are water and salt. For instance, HCl and NaOH react as follows to yield water and sodium chloride:

$$HCl + NaOH \rightarrow NaCl + H_2O$$

Salts dissociate entirely in aqueous solutions. For example:

$$NaCl\ (s) \rightarrow Na^+\ (aq) + Cl^-\ (aq)$$

$$NaClO\ (s) \rightarrow Na^+\ (aq) + ClO^-\ (aq)$$

A solution that resists a pH change (maintaining a relatively constant pH) after small additions of an acid or a base, is called a **buffer**. It is prepared by mixing both a **weak acid and its salt** or both **a weak base and its salt** in water. Note that the salt of a strong acid and a strong base (e.g., NaCl from NaOH and HCl, respectively) cannot be used. Buffers are effective only when small additions of an acid or a base are made. Significant additions of either will destroy the buffer.

There are many naturally occurring buffers. For example, human blood is buffered with a mixture of the anions of two salts, bicarbonate and carbonate: HCO_3^- and CO_3^{-2}. Buffers are used in medications, such as in buffered aspirin or eye wash as well as in swimming pools to prevent the growth of bacteria and algae.

84 Experiment 11 Titration of Buffers

Safety Tips

Review from the first class:

1. Wearing goggles
2. Handling liquid and solid chemicals
3. Handling solutions of chemicals
4. Special case: solutions of acids and bases

Follow any additional instructions given by your professor in the:

1. Use and care for the burette
2. Operation and care for pH meter (or pH sensor of a computer interface).

Computer Procedure

a) Calibration of pH Sensor

1. Calibrate your pH sensor according to the calibration procedure that follows.
   a) Obtain a small amount of pH buffers and place in a small beaker. In principle the buffers can be of any two pH values but the range 4 and 9 is preferred.
   b) On the interface unit itself, have your pH sensor connected to a BNC receptor. Place the sensor into the beaker with the lower pH buffer, making sure that the tip of the sensor is covered by the buffer solution.
   c) On the sensor calibration menu, click on the pH.
   d) Once the pH sensor window opens, look at the sensor voltage and wait until the mill volt reading stabilizes. At this point enter into the space marked "enter the pH" the pH value of your buffer.
   e) Remember to rinse the pH sensor with distilled water when changing from one solution to the next.
   f) Place the pH sensor into the next higher pH buffer and record its value.
   g) The sensor is now calibrated.
2. After the calibration procedure is complete, set up the program to plot pH on the y-axis and keyboard entry on the x axis.

b) Effect of HCl Addition to Distilled Water

3. Place in a 125 mL Erlenmeyer flask 40.0 mL of the 0.2 M weak acid and 20.0 mL of 0.2 M NaOH. Swirl the mixture thoroughly. **Label this solution as buffer #1.**
4. Place in a second 125 mL Erlenmeyer flask 40.0 mL of the 0.2 M weak acid and 30.0 mL of 0.2 M NaOH. Swirl the mixture thoroughly. **Label this solution as Buffer #2.**
5. Clamp a burette. **Use a short stem funnel** to carefully fill it with 0.1 M HCl.
6. Use a graduated cylinder to transfer 100 mL distilled water in a 250 mL beaker and place it on the stirrer platform. Gently insert (**do not drop!**) a stirring bar in the beaker. Turn on the stirrer slowly.

7. Clamp the pH sensor to a ring stand and carefully lower it into the beaker, making sure that the sensor does not touch the stirring bar.
8. Place the tip of the clamped filled burette (from step 5) over the beaker.
9. You are now ready to start collecting data.
10. A window will open, asking you to enter a value. The first value will be 0. Click "OK" and the window will open again.
11. Add 1.0 mL HCl from the burette into the beaker containing the stirring bar. Enter 1.0 mL on the computer screen. Wait 1–2 seconds and then click "OK."
12. Repeat step 11 by the successive addition of 2.0 mL HCl at a time until a total of 20.0 mL of HCl has been added. After **every 2.0 mL HCl addition** enter the volume added to the window space on the computer screen.
13. Print your graph.
14. Remove the pH sensor from the beaker and insert in another beaker that contains only distilled water.

c) Effect of HCl Addition to Buffer #1

15. Use a graduated cylinder to transfer 100 mL distilled water in a 250 mL beaker and place it on the stirrer platform of the stirrer. Gently insert (**do not drop!**) the stirring bar in the beaker. Turn on the stirrer slowly.
16. Use a graduated cylinder to add exactly 25 mL of **buffer #1** to the beaker in step 15.
17. Remove the pH sensor from the beaker containing distilled water (step 14) and clamp to the ring stand. Carefully lower it into the beaker, making sure that the sensor does not touch the stirring bar.
18. Place the tip of the clamped burette (from step 17) over the beaker. **You may have to refill it before you continue with step #19.**
19. You are now ready to start collecting data.
20. A window will open, asking you to enter a value. The first value will be 0. Click "OK" and the window will open again.
21. Add 1.0 mL HCl on the burette into the beaker containing the stirring bar. Enter 1.0 mL on the computer screen. Wait 1–2 seconds and then click "OK.
22. Repeat step 21 by the successive addition of 2.0 mL HCl at a time until a total of 20.0 mL of HCl has been added. After **every 2.0 mL HCl addition,** enter the volume added to the window space on the computer screen.
23. Print your graph.
24. Remove the pH sensor from the beaker and insert in another beaker that contains only distilled water.

d) Effect of HCl Addition to buffer #2

25. Repeat steps 15 through 24 using **buffer #2** instead of buffer #1 in step 16.
26. Drain the remaining HCl from the burette. Discard the remaining HCl in the waste container as directed by your instructor.
27. Use a short stem funnel to fill the burette with water and drain again. Discard the water.

e) Effect of NaOH Addition to Distilled Water

28. Use the short stem funnel to fill the burette with 0.1 M NaOH.
29. Repeat steps 6 through 14 using 0.1 M NaOH instead of 0.1 M HCl.

f) Effect of NaOH Addition to Buffer #1

30. Repeat steps 15 through 24 using 0.1 M NaOH instead of HCl.

g) Effect of NaOH Addition to Buffer #2

31. Repeat steps 25 through 27 using 0.1 M NaOH instead of HCl.
32. Drain the burette and place the remaining NaOH in the designated waste container following your instructor's directions.
33. Use a short stem funnel to fill the burette with distilled water. Drain the burette and discard the water. Return the burette to the cart.

Figure 11.1 Titration set up

Date: _____

Class: _____

Name: _____

Experiment 11—Data Sheet

Titration of Buffers

Table 11.1 Computer Procedure

Volume Added (mL)	0.1 M HCl to Buffer pH	0.1 M HCl to Water pH	0.1 M NaOH to Buffer pH	0.1 M NaOH to Water pH
V = 0.0	_____	_____	_____	_____
V = 1.0	_____	_____	_____	_____
V = 2.0	_____	_____	_____	_____
V = 3.0	_____	_____	_____	_____
V = 4.0	_____	_____	_____	_____
V = 5.0	_____	_____	_____	_____
V = 6.0	_____	_____	_____	_____
V = 7.0	_____	_____	_____	_____
V = 8.0	_____	_____	_____	_____
V = 9.0	_____	_____	_____	_____
V = 10.0	_____	_____	_____	_____
V = 11.0	_____	_____	_____	_____
V = 12.0	_____	_____	_____	_____
V = 13.0	_____	_____	_____	_____
V = 14.0	_____	_____	_____	_____
V = 15.0	_____	_____	_____	_____

Table 11.2 Actual Hands-on Procedure

Volume Added (mL)	0.1 M HCl to Buffer pH	0.1 M HCl to Water pH	0.1 M NaOH to Buffer pH	0.1 M NaOH to Water pH
V = 0.0	_____	_____	_____	_____
V = 1.0	_____	_____	_____	_____
V = 2.0	_____	_____	_____	_____
V = 3.0	_____	_____	_____	_____
V = 4.0	_____	_____	_____	_____
V = 5.0	_____	_____	_____	_____
V = 6.0	_____	_____	_____	_____
V = 7.0	_____	_____	_____	_____
V = 8.0	_____	_____	_____	_____
V = 9.0	_____	_____	_____	_____
V = 10.0	_____	_____	_____	_____
V = 11.0	_____	_____	_____	_____
V = 12.0	_____	_____	_____	_____
V = 13.0	_____	_____	_____	_____
V = 14.0	_____	_____	_____	_____
V = 15.0	_____	_____	_____	_____

Questions and Problems

I. Which of the following solutions will act as a buffer?
   1. HNO_3 and KNO_3
   2. HNO_2 and KNO_2
   3. H_2SO_4 and Na_2SO_4
   4. H_2SO_3 and Na_2SO_3
   5. Acetic acid and sodium acetate
   6. Ammonia and ammonium chloride

II. Give the conjugate base for each of the following acids:
   1. HNO_2

   2. H_2SO_3

   3. HCl

   4. HCN

Experiment 12

Nuclear Chemistry: Radioactivity

Introduction

Isotopes of an element are species that have the same number of protons but a different number of neutrons. Some isotopes are unstable and decay with time by emitting radiation (particles and/or energy). This property is referred to as **radioactivity**, and these isotopes are referred to as **radioactive isotopes**. Since these processes involve the nuclei of atoms, the area of chemistry that studies radioactivity is called **nuclear chemistry**.

Three of the most common types of radiation are designated by the first three letters of the Greek alphabet: **alpha, beta,** and **gamma**. Each **alpha** particle is composed of two protons and two neutrons, which happens to be identical to the nucleus of a helium (He) atom. **Beta** particles are electrons that are generated in the nuclei of isotopes, while **gamma** rays are part of the spectrum of electromagnetic radiation. Since there are many other sources of radiation, such as television and electric power lines, a background radiation always exists. Radiation may be damaging to living organisms when the dosage is high.

This experiment will investigate the effect of increasing the distance from the source and the use of protective shielding on the amount of radiation received. Background radiation will be measured prior to and subtracted from any measurement. The dosage of radiation emitted by the used source is insignificant and less than the one produced by a color TV. The instrument that detects radiation is a Geiger counter and counts photons or electrons that strike its sensor and records radiation as **counts per minute** (cpm).

Safety Tips

Follow your professor's instructions on the:

1. Use of the Geiger counter
2. Use of sources of radiation

I. Radiation vs Distance

Procedure

1. Turn on the Geiger counter 5 minutes prior to the beginning of the experiment.
2. Count the background radiation for 1 minute and record it on the data sheet.
3. Set the count interval for 1 minute, for the rest of the experiment.
4. Place the β radiation source with the label side up on the plastic plate with a smaller hole, and insert the plate into the first groove from the top of the sensor box (1 cm = 10 mm distance from the sensor).
5. Count the radiation and record its value cpm in Table 12.1 of the data sheet.
6. Move the plate with the β radiation source to the next groove from top of the sensor box (corresponding to 2 cm = 20 mm distance from the sensor).
7. Repeat step 5.
8. Repeat step 5 for the 3^d, 4th, 5th, 6th grooves (corresponding to 3 cm = 30 mm, 4 cm = 40 mm, 5 cm = 50 mm, 6 cm = 60 mm from the sensor, respectively).

Tabulating

1. Subtract the background radiation from each cpm value recorded in Table 12.1 to obtain the net cpm for each groove.
2. Record the results in Table 12.1.

Plotting

Plot 1 Radiation vs Distance
Construct a graph by plotting net cpm (y-axis) vs distance in mm (x-axis) to obtain a smooth curve. Note that the high values of radiation on the y-axis correspond to the low values of the distances on the x-axis.

Plot 2 Radiation vs the Reciprocal of the Distance Squared
(Inverse Square Law)
Use the same data to plot net cpm (y-axis) vs $1/\text{distance}^2$ (x-axis). Note that the values on the y-axis are the same as in Plot 1 (net cpm). The values of distance are squared and the inverse used as values on the x-axis. The graph should be a straight line.

II. Radiation vs Thickness of the Shielding

Procedure

1. Place the β-radiation source on the plastic plate with a smaller hole and insert the plate into groove #4 from the top of the sensor box. **The source will stay in this groove for the rest of this part of the experiment.**
2. In a way similar to Part I, collect the measurements as radiation without shielding for the 40 mm distance.
3. Record the data in Table 12.2A as counts per minute (cpm).
4. Place the thinnest polyethylene shielding on the other plate (with the wider hole), and insert it into the second groove from the top of the sensor box. Keep the source where it was placed in step 1, in groove #4.

5. Count the radiation value and record it in Table 12.2A.
6. Replace the shielding with the one that is next in increasing thickness.
7. Repeat step 5.
8. Repeat steps 6–7 for the rest of the available shielding.

Tabulating

1. Subtract the background radiation from each radiation value (cpm) in Table 12.2A.
2. Record the new values as net cpm in Table 12.2A.

Plotting

Construct a graph by plotting net cpm (y-axis) vs thickness of the shielding (x-axis) to obtain a smooth curve. Note that the high values of radiation on the y-axis correspond to thinner shielding on the x-axis, and the low values of radiation correspond to thicker shielding.

III: Radiation vs Shielding Material

Repeat Part II for aluminum shielding, and recording data in Table 12.2B. Then repeat Part II for lead shielding, recording data in Table 12.2C. Finally, compare Tables 12.2A, 12.2B, and 12.2C to see how radiation is effected by the nature of the blocking material.

IV: Computer Simulation Experiment "Nuclear Chemistry: Half Life of a Radionucleotide"

Perform the computer simulation experiment "Nuclear Chemistry: Half life of a Radionucleotide" using "Bridge to the Lab" software.

Date: Nov. 30, 2010
Class:
Name: Kevin Sharma

Experiment 12—Data Sheet

Nuclear Chemistry: Radioactivity

Background radiation: __11__ cpm

Table 12.1 Radiation vs Distance

Groove #	Distance (d)	d²	1/d²	cpm	Net cpm *
1	10 mm	100	1/100	709	698
2	20 mm	400	1/400	1616	1605
3	30 mm	900	1/900	734	723
4	40 mm	1600	1/1600	474	463
5	50 mm	2500	1/2500	288	277
6	60 mm	3600	1/3600	184	173

*Net cpm = cpm−background

Table 12.2A Radiation vs Shielding Polyethylene Thickness

Shielding	Shielding Thickness (mm)	Cpm	Net cpm*
1	14.1 mm	709	698
2	28.1 mm	676	665
3			
4			

*Net cpm = cpm−background

Table 12.2B Radiation vs Aluminum Thickness of the Shielding

Shielding	Shielding Thickness (mm)	Cpm	Net cpm*
1	161	399	398
2	328	150	139
3	590	39	28
4			

*Net cpm = cpm−background

Table 12.2C Radiation vs Lead Thickness of the Shielding

Shielding	Shielding Thickness (mm)	Cpm	Net cpm*
1	1230	14	4
2	1890	14	1
3	3632	23	0
4	7435	12	−2

*Net cpm = cpm−background

Questions and Problems

1. Does the amount of radiation (cpm) increase or decrease as distance between the source and the sensor increases?

2. Does the amount of radiation (cpm) increase or decrease as thickness of the shielding increases while the distance between the source and the sensor stays the same?

3. Does the amount of radiation (cpm) change when shielding made from different material is used?

4. Based on answers 1–3, what are three factors that allow us to reduce the amount of radiation?

Experiment 13

Find (handwritten)

Qualitative Analysis

$10^{-2} = .01$ (handwritten)

Introduction

There are many chemical reactions in aqueous solutions, which are accompanied by a visual change, such as the appearance of a precipitate, evolution of gas, or color change. Many of these reactions involve ions, which are dissolved in water.

Often one of the ions can be a suspected contaminant (**pollutant**) in water and its presence may be identified by its reaction with another ion of opposite charge that may be added by the analyst. The chemical reaction and the visual change between them is referred to as a **test** for the pollutant and the ion from the laboratory shelf that is added by the analyst is referred to as a **reagent** for the pollutant. If the pollutant ion is present in the water sample, the chemical reaction will take place. If the pollutant ion is not present in the water sample, there will be no chemical reaction and thus no visual change.

In most cases, a sample of air, water, or soil may contain more than one pollutant. As a result the suspected pollutants have to be separated from each other into separate test tubes by various separation techniques first. These techniques may include, among others, chromatography or precipitation. Then the identification of each pollutant is achieved by performing the tests described previously. These tests are the vehicles through which the **qualitative analysis** of an ion is achieved.

In this laboratory experiment the identification tests for positive pollutant ions will be studied using negative ions that will serve as the reagents. All such reactions will be performed in aqueous solutions. The cations will be Pb^{2+}, Hg_2^{2+}, Cu^{2+}, Ag^+, Zn^{2+}, which will be present in solution as the corresponding nitrate salts. These cations will be identified by the visual changes that occur when aqueous solutions of HCl, H_2S, K_2CrO_4, HNO_3, and NH_4OH will be added as reagents.

This experiment will be performed using a computer program called "Bridging to the Lab."

Safety Tips

Review from the first class:

1. Wearing goggles
2. Handling liquid and solid chemicals as well as solutions

Procedure

I. Computer Simulation Experiment

1. Turn your computer on and log in.
2. Open the "Bridging to the Lab"* program written by Loretta Jones and Roy Tasker.
3. From the main menu pick "Inorganic Qualitative Analysis: Identifying the Metal Ions."
4. The first window will display the objectives and a list of all the sections.
5. Move the cursor over "What You Should Know" (on the lower left-hand corner) to see what you need to be familiar with in order to perform this simulation.
6. When you are ready, click the "Start" button (on the lower right-hand corner).
7. Follow the instructions and perform all tasks as indicated. Click the "Next" button (lower right-hand corner) every time you need to move to the next step. **Note: the program will not let you move on until you correctly perform all tasks called for in that section. You must carefully record the reagent addition sequence on the flow chart, as you will need it to complete the second part of this experiment.**
8. As you are performing the experiment fill in the flow chart that appears on the adjacent page of this lab manual.
9. Complete all sections: **Introduction, Experimental Setup, Collecting Data, Summary,** and **Self-Test.**
10. The last window will summarize all the sections completed. Show this to your instructor before moving on.
11. Close the program and shut down the computer. Proceed with Part II.

II. Actual Qualitative Analysis Experiment

1. Select an unknown solution and record its number.
2. Use the same procedure you used in the computer part of the experiment to identify this unknown. Follow the same sequence of reagent addition and carefully **record your observations on the data sheet.**

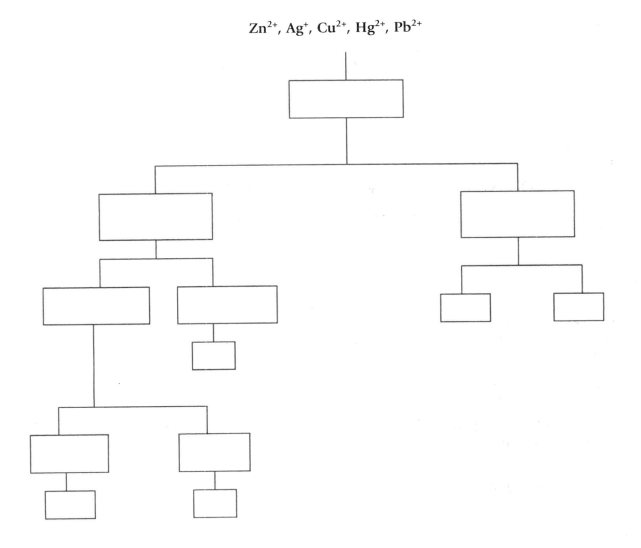

Flow Chart that summarizes the reagent addition sequence using the "Bridging to the Lab"* program written by Loretta Jones and Roy Tasker.

Date: 4/27/10
Class: CH-127
Name: Kevin Sharma

Experiment 13—Data Sheet
Qualitative Analysis

Data Sheet

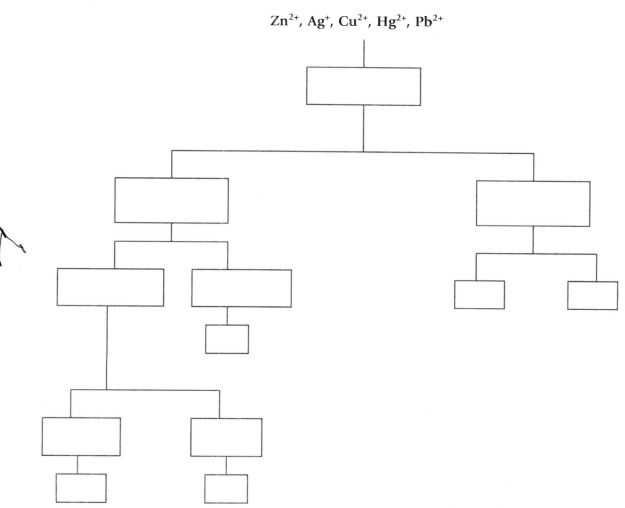

Unknown number: 24   My unknown cation is: _____

Experiment 14

Chemical Equilibrium and Le Chatelier's Principle

Introduction

In any chemical reaction the substances that appear on the left-hand side are called **reactants,** and those that appear on the right-hand side are called **products.** Thus in the reaction:

$$aA + bB \rightleftharpoons cC + dD$$

A and B are the reactants and C and D are the products.

Chemical reactions never go to completion in either direction. Instead they reach a certain state of balance, called **equilibrium,** which is expressed as the ratio of concentrations of the products over the reactants. The value of this expression is unique for a specific reaction under a specific set of conditions of temperature and pressure and is called **equilibrium constant, K_{eq}**. The general form of the equilibrium constant for the previous reaction is given by:

$$K_{eq} = \frac{[C]^c \times [D]^d}{[A]^a \times [B]^b}$$

For reactions that involve solutions the concentrations are expressed in **molarity** units, M, or moles of solute per liter of solution. For gas reactions, partial pressure of gases (in atm units) may also be used.

The position of equilibrium (i.e., the ratio of concentrations of products over reactants) is different and unique for each chemical equilibrium reaction. For example, if one starts with a certain quantity for each of the reactants and no products, the reaction proceeds in the forward direction to consume part of the starting quantities of the reactants and produce some of the products. On the other hand, if one starts with some product quantities and no reactants, the reaction will proceed in the reverse direction in a way that will establish the K_{eq} again.

As soon as a change is imposed on a system at equilibrium, the reaction shifts in a response that will reestablish the equilibrium ratio. This is known as **Le Chatelier's principle.** Thus,

a) if one or more products is/are removed, in the form of a precipitate or a gas, more of the reactants will be consumed so as to establish the equilibrium constant value.
b) If one of the reactants is added in excess, more products will be formed in order to achieve the equilibrium constant.

c) If the reaction is exothermic, adding heat will reverse the reaction. On the other side endothermic reactions will yield larger quantities of the product(s) when heat is applied.

In this experiment two chemical equilibrium reactions in solution will be studied. By manipulating the concentrations of the reactants, the equilibrium will be perturbed and the shift of the reaction system to reestablish equilibrium will be visually observed as a color change. A third chemical reaction will be used to study the effect of temperature change on the position of equilibrium.

> ### Safety Tips
> Review from the first class:
> 1. Wearing goggles
> 2. Handling liquid and solid chemicals as well as their solutions

a) Reaction of Fe^{3+} and SCN^-

The forward reaction involves the yellow ferric (Fe^{3+}) cation and the colorless thiocyanate anion to form the deep-red complex ferric thiocyanate ($FeSCN^{2+}$) ion. The reverse reaction involves the decomposition of the ferric thiocyanate ion into the ferric cation and thiocyanate anion.

$$Fe^{3+}(aq) + SCN^-(aq) \leftrightarrows FeSCN^{2+}(aq)$$
$$\text{yellow solution} \quad \text{colorless solution} \quad \text{deep red solution}$$

The source of the ferric cations is ferric nitrate ($Fe(NO_3)_3$) in solution and the source of the thiocyanate ions is sodium thiocyanate (NaSCN) in solution. The complete reaction is:

$$Fe(NO_3)_3\ (aq) + NaSCN\ (aq) \leftrightarrows FeSCN^{2+}\ (aq) + NaNO_3\ (aq) + 2\ NO_3^-$$

Procedure

1. Place 50 mL of distilled water into either a 150 mL or 100 mL beaker.
2. Add 10 drops of 0.1 M stock solution of ferric nitrate and 10 drops of 0.1 M stock solution of sodium thiocyanate to the beaker from step 1. Swirl the mixture to allow for the equilibrium to be established. This will be your reference solution.
3. Label four clean and dry small test tubes as A, B, C, and D.
4. Place 10 drops of the reference solution into each test tube from step 3.
5. Test tube A will serve as the reference. The contents of test tubes B, C, and D will be compared to those of test tube A after the equilibrium is perturbed in them.
6. Add 5 drops of 0.1 M stock solution of ferric nitrate (a reactant) to test tube B. Swirl the contents of test tube B and record your observations.

7. Add 5 drops of 0.1 M stock solution of sodium thiocyanate (another reactant) to test tube C. Swirl the contents of test tube C and record your observations.
8. Add 5 drops of 0.1 M stock solution of sodium nitrate (the other product) to test tube D. Swirl the contents of test tube D and record your observations.

b) Reaction of Cu^{2+} and aqueous NH_3

The forward reaction is between the colorless aqueous ammonia, NH_3, and the turquoise blue Cu^{2+} cation leading to the formation of the deep royal blue copper-ammonia $Cu(NH_3)_4^{2+}$ complex ion. The reverse reaction is the decomposition of the deep blue complex copper-ammonia ion into the colorless aqueous ammonia and turquoise blue copper (II) cation.

$$Cu^{2+}(aq) + 4NH_3(aq) \leftrightarrows Cu(NH_3)_4^{2+}(aq)$$
turquoise blue solution   colorless solution   deep royal blue solution

The source of the copper (II) cation is an aqueous solution of copper sulfate, and the source of ammonia is the aqueous ammonium hydroxide solution. So, the reaction can be written as:

$$CuSO_4(aq) + 4NH_3(aq) \leftrightarrows Cu(NH_3)_4SO_4(aq) + H_2O(l)$$

The experiment will involve the addition of aqueous ammonia and its removal. The removal will be achieved by neutralizing ammonia with hydrochloric acid (HCl).

Procedure

1. Place 10 drops of 0.1 M stock solution of copper sulfate into a test tube.
2. Add dropwise 1 M aqueous ammonia solution (ammonium hydroxide) with stirring to the test tube from step 1, until the turquoise blue color changes into deep royal blue. Record the number of drops in the data sheet.
3. Add dropwise 1 M hydrochloric acid (HCl) solution with stirring to the contents of the test tube from step 2, until the deep royal blue color changes into turquoise blue. Record the number of drops in the data sheet.

c) Equilibrium between Tetracoordinated and Hexacoordinated Co^{2+}

This section deals with the temperature dependence of K_{eq}, for the chemical reaction shown here:

$$Co(H_2O)_6^{2+}(aq) + 4Cl^-(aq) \leftrightarrows CoCl_4^{2-}(aq) + 6H_2O(l)$$
pink complex   purple/blue complex

At 5 °C, aqueous Co^{2+} ions are present in an aqueous solution of the coordinated complex $Co(H_2O)_6^{2+}$(aq), cobalt hexahydrate. If there is an excess of Cl^- ions present in the same solution, then in addition to the above complex, four Cl^-(aq) ions are coordinated with Co^{2+}(aq) producing the complex ion $CoCl_4^{2-}$(aq), cobalt (II) tetrachloride. Both complex ions are at equilibrium with each other and K_{eq} (i.e., the ratio of the relative concentrations) is temperature dependent.

The position of the equilibrium can be shifted in either direction by changing the temperature. The decrease of temperature shifts the equilibrium to the left—that is, as the temperature decreases, the concentration of the more stable cobalt hexahydrate (pink) complex increases while the concentration of the less stable cobalt tetrachloride (blue) complex decreases. Increasing the temperature shifts the equilibrium to the right favoring the formation of the less stable blue cobalt (II) tetrachloride complex.

The concentration of a colored species in a solution is manifested as the intensity of its color. The change in the concentration of that species leads to the change in the intensity of its color. Thus, the higher the concentration, the greater the intensity of the color and vice versa.

The source of Co^{2+} (aq) is the salt $CoCl_2$ (s), which dissociates in water into Co^{2+}(aq) and Cl^-(aq). There will be two sources of Cl^- (aq); one from the ions produced by the ionization of the $CoCl_2$ (s) and the other one being an aqueous solution of HCl that is added to produce an excess of Cl^- (aq). The aqueous solutions will be the source of water molecules.

Procedure

1. Place 20 drops of the stock solution containing both complexes into each of three different small test tubes labelled A, B, and C. The first test tube (A) will be heated, the second test tube (B) will be cooled, and the third one (C) will be used as a reference.
2. Observe and record in the data sheet the initial color of all three test tubes.
3. Place test tube (A) into a water bath and start heating slowly. Observe the color change of the test tube contents as the heating increases.
4. Stop heating when the temperature of the water bath reaches 90°C.
5. Remove test tube A from the hot water bath and place it next to the reference test tube (C). Record and compare the color of the hot test tube A contents with those of test tube C on the data sheet.
6. Observe the color change as the test tube is cooling off. Record the color of the test tube A contents when they cool to room temperature.
7. Place test tube B into an ice-water bath and observe the color changing.
8. Take the test tube out of the ice-bath when the temperature is about 5°C and place it next to the reference test tube C. Record and compare the color of the cold test tube B contents with those of test tube C on the data sheet.
9. Allow the contents of test tube B to warm to room temperature. Record the color change.

d) Computer Simulation Experiment

The students will perform the computer simulation experiment "Equilibrium: the Chemistry of Magic" using "Bridging to the Lab" software to investigate the dynamic nature of equilibrium processes in general and chemical reactions, in particular.

Date: 5/4/10
Class: CH-Lab
Name: Kevin Sharma

Experiment 14—Data Sheet

Chemical Equilibrium and Le Chatelier's Principle

Observations and Conclusions

a) Reaction of Fe^{3+} and SCN^-

Test Tubs #	Action Performed	Observation	Conclusion
A	XXX	XXX	XXX
B	Adding $Fe(NO_3)_3$ (aq)		
C	Adding NaSCN (aq)		
D	Adding $NaNO_3$ (aq)		

b) Reaction of Cu^{2+} and Aqueous NH_3

Action Performed	Observation	Conclusion
Mixing $CuSO_4$ and aqueous NH_3		
Adding aqueous NH_3		
Adding aqueous HCl		

c) Equilibrium between Tetracoordinated and Hexacoordinated Co^{2+}

Test Tube #	Action Performed	Observation	Conclusion
A	Heating the test tube to 95°C		
B	Cooling the test tube to 5°C		
C	Dissolving $CoCl_2$ in water at room temperature		

Questions and Problems

(Show all calculations)

1. Consider the endothermic reaction:

   C_2H_6O (l) + $C_4H_8O_2$ (l) ⇌ $C_6H_{12}O_2$ (l) + H_2O (l)

   What will be the effect of the following on the equilibrium—that is, will the reaction be shifted to the left, right, or there will be no effect?
   a) Adding excess C_2H_6O to the reaction mixture?

   b) Performing the reaction using wet glassware?

   c) Removing $C_6H_{12}O_2$ as soon as it is formed?

   d) Raising the temperature?